# RESILIENCE ENGINEERING PERSPECTIVES

# Resilience Engineering Perspectives
## Volume 2
### Preparation and Restoration

Edited by

CHRISTOPHER P. NEMETH

*Klein Associates Division (KAD) of Applied Research Associates (ARA), USA*

ERIK HOLLNAGEL

*Professor, MINES ParisTech, France*

&

SIDNEY DEKKER

*Professor, Lund University School of Aviation, Sweden*

**CRC Press**
Taylor & Francis Group
Boca Raton  London  New York

CRC Press is an imprint of the
Taylor & Francis Group, an **informa** business

CRC Press
Taylor & Francis Group
6000 Broken Sound Parkway NW, Suite 300
Boca Raton, FL 33487-2742

First issued in paperback 2019

© 2009 by Christopher P. Nemeth, Erik Hollnagel and Sidney Dekker
CRC Press is an imprint of Taylor & Francis Group, an Informa business

No claim to original U.S. Government works

ISBN-13: 978-0-7546-7520-4 (hbk)
ISBN-13: 978-0-367-38540-8 (pbk)

International Standard Book Number-13: 978-0-7546-7520-4 (Hardback)

**Visit the Taylor & Francis Web site at**
**http://www.taylorandfrancis.com**

**and the CRC Press Web site at**
**http://www.crcpress.com**

# Contents

# List of Figures

# List of Tables

# About the Contributors

Theodore T. Allen is an Associate Professor of Industrial and Systems Engineering at The Ohio State University. He is the author of over 35 refereed publications including a textbook on engineering statistics. He uses mathematical techniques for the design of experiments to solve problems in information search, bioinformatics, and voting machine allocation. Dr Allen is a senior member and officer of the American Society of Quality and an officer of two sections of INFORMS. He is an associate editor of the *Journal of Manufacturing Systems*. Dr Allen is also a nationally recognized expert on improving voting systems. Email: allen.515@osu.edu

Tammy E. Beck, PhD, is an Assistant Professor of Management in the Belk College of Business at the University of North Carolina at Charlotte. She received her PhD in organization and management studies from the University of Texas at San Antonio. Her research interests include organization resilience, the role of middle managers in organizational interpretation, and inter-organization collaboration. In regards to inter-organization collaboration, Dr Beck's research focuses on the sharing of knowledge across organizational boundaries and factors associated with effective collaboration among temporary groups of organizations. Email: tbeck8@uncc.edu

Charlie E. Billings is affiliated with the Cognitive Systems Engineering Laboratory at The Ohio State University. He is a Clinical Professor Emeritus in Preventive Medicine at The Ohio State University and a Fellow of the Aerospace Medical Association and the Royal Aeronautical Society. Prior to returning to the University, he was chief scientist at the NASA-Ames Research Center, where in 1975 he and colleagues developed and implemented the NASA Aviation Safety Reporting System (ASRS). Email: chasbill@ix.netcom.com

Thomas A. Birkland, PhD, is the William T. Kretzer Professor of Public Policy in the School of Public and International Affairs at North Carolina State University. His research focuses on theories of the public policy process, applied to natural disasters and industrial accidents. Before joining NCSU, he was on the faculty at the State University of New York at Albany, where he directed the Center for Policy Research. In 2006, he served as a program officer for the Infrastructure Management and Hazard Response Program at the National Science Foundation. His most recent book, *Lessons of Disaster*, was published by Georgetown University Press in 2006. Email: tom_birkland@ncsu.edu

Richard I. Cook, MD is a physician, educator, and researcher at the University of Chicago. His current research interests include the study of human error, the role of technology in human expert performance, and patient safety. He is internationally recognized as a leading expert on medical accidents, complex system failures, and human performance at the sharp end of these systems. He has investigated a variety of problems in such diverse areas as urban mass transportation, semiconductor manufacturing, and military software systems. He is often a consultant for not-for-profit organizations, government agencies, and academic groups. Dr Cook's most often cited publications are *Gaps in the Continuity of Patient Care and Progress in Patient Safety* (2000), *Operating at the Sharp End: The Complexity of Human Error* (1994), *Adapting to New Technology in the Operating Room* (1996), and the report "A Tale of Two Stories: Contrasting Views of Patient Safety" (1998). Email: ri-cook@uchicago.edu

Sidney Dekker received his PhD from The Ohio State University in 1996 and is Professor of Human Factors and System Safety at Lund University in Sweden. He has previously lived and worked in the USA, Australia, New Zealand, the Netherlands, Singapore and England. His recent books include *Ten Questions about Human Error: A New View of Human Factors and System Safety* (2005), *The Field Guide to Understanding Human Error* (2006), and *Just Culture:*

*Balancing Safety and Accountability* (2007). Email: sidney.dekker@ tfhs.lu.se

Erik Hollnagel (PhD, psychology) is Professor and Industrial Safety Chair at MINES ParisTech (France), Professor Emeritus at University of Linköping (Sweden), and Visiting Professor at the Norwegian University of Science and Technology (NTNU) in Trondheim (Norway). Since 1971 he has worked at universities, research centres, and industries in several countries and with problems from several domains, including nuclear power generation, aerospace and aviation, air traffic management, software engineering, healthcare, and land-based traffic. His professional interests include industrial safety, resilience engineering, accident investigation, cognitive systems engineering and cognitive ergonomics. He has published more than 250 papers and authored or edited 13 books, some of the most recent titles being *Resilience Engineering Perspectives: Remaining Sensitive to the Possibility of Failure* (Ashgate, 2008), *Resilience Engineering: Concepts and Precepts* (Ashgate, 2006), *Joint Cognitive Systems: Foundations of Cognitive Systems Engineering* (Taylor & Francis, 2005) and *Barriers and Accident Prevention* (Ashgate, 2004). Erik Hollnagel is, together with Pietro C. Cacciabue, Editor-in-Chief of the international *Journal of Cognition, Technology & Work*. Email: erik.hollnagel@crc.ensmp.fr

Yao Hu is a PhD student in the Information Systems Department at New Jersey Institute of Technology in Newark, NJ. He received a BE in Automatic Control from Shanghai Jiao Tong University in China and a MS in Computer Science from University of Toledo, Ohio. His current research focuses on cognitive modelling of group decision-making in emergency response. He is a member of the Cognitive Science Society. Email: yao.hu@njit.edu

Cynthia A. Lengnick-Hall, PhD, is Professor of Management at the University of Texas at San Antonio. Her current research projects include studies on organizational resilience; strategic human resource management; managing knowledge for competitive advantage; corporate governance; and developing new perspectives on strategic dilemmas such as the causal ambiguity paradox, competing forces of strategic agility and strategic momentum, and achieving synergy between organizational learning and organizational unlearning. Her work has been published in journals such as the *Academy of Management Review, Academy of Management Journal, Strategic Management Journal, Journal of Management, Human Resource Management*, and *Academy of Management Executive.* She has co-authored three books and the most recent, co-authored with Mark Lengnick-Hall, is *Human Resource Management in the Knowledge Economy: New Challenges, New Roles, New Capabilities.* Email: cynthia.lengnickhall@utsa. edu

David Mendonça, PhD, is an Associate Professor in the Information Systems Department at New Jersey Institute of Technology in Newark, NJ. His research is concerned with modelling and supporting decision-making by individuals and organizations, particularly in emergency situations. This work has involved field studies with organizations such as the US Federal Emergency Management Agency, the US Army Corps of Engineers, the Port of Rotterdam (The Netherlands), Consolidated Edison and Verizon. His work has been supported by a number of grants from the US National Science Foundation, including a CAREER award in 2005 and three awards pertaining to the 2001 World Trade Center attack. He is a member of the IEEE, Institute for Operations Research and Management Sciences, and the Cognitive Science Society. His research has been published in a number of journals including *Decision Support Systems, IEEE Transactions on Systems, Man and Cybernetics (Part A)*, and the *European Journal of Operational Research.* Email: david.mendonca@njit.edu

Jordan Multer is manager of the rail human factors program at the Volpe National Transportation Systems Center. He has evaluated the impact of technology on human performance in aviation, marine and railroad operations. He is currently implementing a confidential close-call reporting system for the US railroad industry as part of a Federal Railroad Administration imitative to develop more proactive methods for managing safety. Email: jordan.multer@dot.gov

Christopher P. Nemeth, PhD, studies human performance in complex high hazard environments as a as a Principal Scientist at Klein Research Associates Division of Applied Research Associates. Recent research interests include technical work in complex high stakes settings, research methods in individual and distributed cognition, and understanding how information technology erodes or enhances system resilience. His design and human factors consulting practice and his corporate career have encompassed a variety of application areas from health care to transportation and manufacturing. His consulting practice has included human factors analysis, expert witness, and product development services. His academic career has included adjunct positions with Northwestern University's McCormick College of Engineering and Applied Sciences (Associate Professor), and Illinois Institute of Technology. He serves on the editorial boards of *Cognition, Technology and Work, IEEE Transactions on Systems, Man and Cybernetics (Part A)*, and *Resilience Engineering Perspectives*. His current texts include *Human Factors Methods for Design* (Taylor and Francis/CRC Press) and *Improving Healthcare Team Communications* (Ashgate). Email: cnemeth@ara.com

Martin Nijhof, BSc, is currently employed by KLM Royal Dutch Airlines and works as a Flight Safety Investigator in their Flight Safety Department. His tasking includes investigating flight operational incidents and identifying and investigating flight safety trends. His specialist fields include human factors, automation and the organizational aspects of incidents and accidents. Before joining the KLM Flight Safety team in 2001, he worked as a ground school simulator instructor in the B737 division of KLM's Flight Crew Training Centre. Prior to that, from 1990 to 1996, he was employed by Fokker Aircraft where his primary function was as a Fokker 100 ground school instructor within their Flight Crew Training and Product Support department. Email: martin.nijhof@klm.com

Michael O'Connor, MD, is a physician, educator and researcher at the University of Chicago. His clinical work is a combination of critical care medicine (where he attends in the Medical, Surgical, Cardiothoracic and Burn ICUs), and operating room anaesthesia, where his activity has been centred on anaesthesia for liver transplantation. His educational activity is centred around his clinical activity. Dr O'Connor is also Director of the Senior Medical Student Selective "Vignettes in Physiology." His clinical research has included new drug development (atracurium, sevoflurane, propofol, etomidate, methylnaltrexone, antithrombin, activated recombinant protein C, linezolid, and several blood substitutes), clinical research in critical care (bedside assessment of autoPEEP, use of propofol as a sedative, management of sedation in critically ill patients), and, recently, patient safety. He has lectured about the social science of accidents in a variety of settings and has participated in a variety of activities in the Cognitive Technologies Laboratory. Email: moc5@airway2.bsd.uchicago.edu

Shawna J. Perry, MD, is Director for Patient Safety System Engineering at Virginia Commonwealth University Health Systems in Richmond, VA, as well as Associate Professor and Associate Chair for the Department of Emergency Medicine. Previously, she was Director of Clinical Operations, Chief of Service and Associate Chair of Emergency Medicine at University of Florida Health Sciences Center in Jacksonville, FL where she provided leadership and management for numerous departmental, hospital and university-based initiatives from large-scale coordination of clinical work to the implementation of new information technology (IT) systems into the clinical setting. Since 1996, Dr Perry's primary research interest has been in patient safety, with a particular interest in human factors and ergonomics, the nature of system failures, transitions in care, the impact of IT upon clinical care, organizational behaviour and teamwork. She is widely published on topics related to patient safety, human factors and ergonomics, naturalistic decision-making, communications and emergency medicine. Email: sperry4@mcvh-vcu.edu

Emilie M. Roth, PhD, is principal scientist of Roth Cognitive Engineering. She has been involved in cognitive systems analysis and design in a variety of domains including nuclear power plant operations, railroad operations, military command and control, medical operating rooms and intelligence analysis. She received her PhD in cognitive psychology from the University of Illinois at Champaign-Urbana. She serves on the editorial board of the journals *Human Factors* and *Le Travail Humain*, and is editor of the *Design of Complex and Joint Cognitive Systems* track of the *Journal of Cognitive Engineering and Decision Making*. Email: eroth@verizon.net

Jason Schenk received his PhD from The Ohio State University in 2008 in Industrial Systems Engineering, where he focused on statistical decision-making in the domains of intelligence analysis and organizational resilience. He currently works as a Senior Engineer for DeVivo Automated Systems Technology, Inc., in Huntsville, AL, where he continues his research on joint cognitive systems. Email: jasonschenk@devivoast.com

Ronald Scott, PhD, is a division scientist in the BBN Technologies Intelligent Distributed Computing Department, Cambridge, MA. His research interests include the application of distributed software technology to the design of human-centred applications. He received his doctorate in mathematics from the University of Chicago. Email: rscott@bbn.com

Philip J. Smith, PhD, is affiliated with the Cognitive Systems Engineering Laboratory at The Ohio State University. He is a Professor in the Industrial and Systems Engineering program, with extensive experience in the design of distributed work systems and decision support tools, including the design of the Post-Operations Evaluation Tool, an analysis system used by the FAA and the airlines to evaluate performance in the US airspace system. Email: smith.131@osu.edu

Amy L. Spencer is affiliated with the Cognitive Systems Engineering Laboratory at The Ohio State University. She is a doctoral student in the Industrial and Systems Engineering program, and also has considerable previous work experience with the design of cognitive tools to support collaborative decision-making in the airspace system. Email: amyspencer@columbus.rr.com

Sarah Waterman is a candidate for a Master of Public Administration (MPA) degree at the University of North Carolina at Chapel Hill, and holds a Bachelor of Arts degree in Public Policy from the State University of New York at Albany. Her primary research interests include the role of governmental relations in emergency management and issues of access and equity in healthcare provision. Email: sjwaterman@gmail.com

Robert L. Wears, MD, MS, is an emergency physician, Professor in the Department of Emergency Medicine at the University of Florida, and Visiting Professor in the Clinical Safety Research Unit at Imperial College London. He serves on the board of directors of the Emergency Medicine Patient Safety Foundation, and on the Editorial Board for *Annals of Emergency Medicine*. He is also on the editorial board of *Human Factors and Ergonomics*, the *Journal of Patient Safety*, and the *International Journal of Risk and Safety in Medicine*. Dr Wears has been an active writer and researcher with interests in technical work studies, joint cognitive systems, and particularly the impact of information technology on safety and resilient performance. His work has been funded by the Agency for Healthcare Research and Quality, the National Patient Safety Foundation, the Emergency Medicine Foundation, the Society for Academic Emergency Medicine, the Army Research Laboratory, and the Florida Agency for Health Care Administration. Email: wears@ufl.edu; r.wears@imperial.ac.uk

Ron Westrum, PhD, is Professor of Sociology at Eastern Michigan University. He is a specialist on the cultures of socio-technical systems, and invented the pathological-bureaucratic-generative distinction. Through his consulting organization Aeroconcept, he has advised many technical organizations, including General Motors and Lockheed Missiles and Space. He frequently speaks at national and international meetings on systems safety and risk. In addition to many papers in journals, he is the author of several books, including *Sidewinder: Creative Missile Design at China Lake*. Email: ronwestrum@aol.com

David D. Woods, PhD, is Professor of Cognitive Systems Engineering and Human Systems Integration at The Ohio State University. He studies how people cope with complexity in time-pressured situations such as critical care medicine, aviation, space missions, intelligence analysis, and crisis management. He designs systems to help people find meaning in large data fields when they are under pressure to diagnose and re-plan. His latest work is to model and measure the adaptive capacities of organizations and distributed systems to determine how they are resilient and if they are becoming too brittle in the face of change. He is past-President of the Human Factors and Ergonomics Society, has a Laurel Award from *Aviation Week & Space Technology*, is an advisor to the Columbia Accident Investigation Board, is a member of several National Academy of Science committees, most recently on Dependable Software, and co-author of four books. Email: woods.2@osu.edu

John Wreathall is President and CEO of John Wreathall & Co., Inc., a small consulting company that specializes in the development and application of techniques and tools to improve safety management, including the development of tools and methods associated with resilience engineering. Mr Wreathall is currently working or has worked with clients in the nuclear power, rail transportation and healthcare industries to develop such methods and tools. These are generally associated with measuring the effectiveness of safety management using concepts now recognized to be part of a resilience approach. He has written over 60 published papers, and has been the invited keynote speaker at conferences related to the development and application of resilience including the First Mercosur (South American Free Trade Association) Conference on Safety and Security in Work and its Environment, Porto Alegre (Brazil, 2004), co-chair of the 2005 International Seminar on Resilience Engineering and Cognitive Systems, Rio de Janeiro (Brazil, 2005) and an invited participant to the First and Second International Symposiums on Resilience Engineering, in Söderköping (Sweden, 2004) and Juan-les-Pins (France, 2006). Email: jwreatha@rrohio.com

# Acknowledgements

The authors who have contributed to this text have been generous with their time and thoughts. Those who are the subjects of study, from many walks of life and kinds of work, have made this publication possible. We offer thanks and appreciation to all.

Collage art was created for this volume by fine artist and educator Jeanine Coupe Ryding, of Evanston, IL, USA.

# Chapter 1

# The Ability to Adapt[1]

Christopher P. Nemeth

Not being able to control events, I control myself; and I adapt myself to them, if they do not adapt themselves to me.

*Michel de Montaigne (1533–1592), French essayist.*
*"Of Presumption," The Essays (Les Essais),*
*bk. II, ch. 17, Abel Langelier, Paris (1588)*
*Source: Andrews et al., 1996*

The first volume in the *Resilience Engineering Perspectives* series (Hollnagel, Nemeth and Dekker, 2008) grew from papers that were presented at the *Second Symposium on Resilience Engineering* at Juan les Pins, France, in November 2006. As the first volume in the series outside of a symposium setting, this text considers what it means for an organization to be able to adapt, and how that ability relates to resilience and efforts to engineer resilient systems. Its authors deliberate on how that ability plays out before and after a critical challenge – the *preparation* to meet challenges and the *restoration* that returns a system to working order afterward.

*Resilience Engineering* (RE) (Hollnagel et al., 2006) focuses on the ability of an organization to cope with, and recover from, unexpected developments. Rather than a focus on a system's

1     Dr Nemeth's research has been supported by the Department of Anesthesia and Critical Care at the University of Chicago, the US Food and Drug Administration, and a grant from the Agency for Healthcare Research and Quality. The author offers his sincere thanks to Erik Hollnagel for his valuable insights and comments during the preparation of this chapter.

productive *capacity*, RE can be used to assess and enhance the ability of an organization to *adapt* in order to meet challenges. With roots in complexity study (Carlson and Doyle, 2002) and cognitive systems engineering (Hollnagel and Woods, 2005), RE seeks to create and maintain systems that can cope with and adapt to complex, dynamic, and changing environments. RE acknowledges the inability to specify all possible threats and responses. Instead, it provides methods and tools to manage safety and productivity.

On the face of it, the ability to adapt has features that most organizations would want. Being able to adjust to different conditions or environments sounds like a desirable trait. It offers the potential for a system to change to new modes of operation, much like the crew of a Navy ship shifting to "General Quarters." The analogy ends there, though. Sending all hands on a ship to their battle stations is a proven, accepted need that has been tested over time. The need and ability for organizations to make substantial changes to prepare for potential threats is more complex. Change poses a potentially unsettling challenge for organizations that seek to maintain the status quo.

Studies of system failure reasonably lead to the question of what to do about it. The easy, but inadequate, response is to react by making a quick revision. The harder, but more substantive, response is to understand the setting in which systems exist and create essential features that have the potential to resolve such issues long term. The following chapters examine the nature and implications of taking the harder, more substantive approach.

## Probing Resilience

The chapters in this text spell out the very real considerations of how resilience and engineering efforts to create resilient systems play out in the real world. They also view the topic from viewpoints that are outside of systems safety and engineering. Notions of resilience have evolved in multiple disciplines, including organizational research, traditional risk management, and complexity studies. Organizational research views resilience as the need for collective mindfulness (Wildavsky, 1988; Weick et al., 1999), and chapters in this volume by Birkland and Lengnick-Hall and Beck provide insightful discussions on this theme.

Traditional risk assessment (Rasmussen, 1983) has approached resilience as a minor variation in performance due to over- or under-adaptation. Complexity studies (Csete and Doyle, 2002) approach resilience as an engineering and ecological issue, and most of the chapters in the volume reflect this perspective.

## Resilience is Not Reliability

Those who are less familiar with the resilience literature may not see the difference between resilience and what has come to be known as high reliability (Rochlin et al., 1987). The initial call for a symposium on resilience engineering explained the difference. "It is not enough that they are reliable so that the failure probability is acceptably low. They must also be resilient and have the ability to recover from irregular variations, disruptions and a degradation of expected working conditions" (Hollnagel et al., 2004). A broader comparison between reliability and resilience might help to explain why this is so.

Few contributions to the notion of the high reliability organization (HRO) rival Rochlin, LaPorte and Roberts' (1998) analysis of US Navy aircraft carrier flight operations. That seminal work identified four aspects of reliability:

1. *Self design and self replication* – Tasks are broken down internally into decomposition rules that are often *ad hoc* and circumstantial. The organization is integrated horizontally, vertically, and across command structures. Structures shift in time to adapt to varying circumstances. Continual training and retraining develops, transmits and maintains the information needed for safe and efficient operation. Workups intensify training prior to operational deployment. Objects, events, situations, and appropriate conduct are codified. Assignments are rotated regularly.

2. *The paradox of high turnover* – Efforts to manage rapid crew turnover benefit the organization. Turnover requires officers to command respect among senior petty officers. Organizational and technical innovation are resisted until proven to benefit operations, then quickly diffused throughout the fleet. Standard operating procedures (SOP) and procedures are unusually robust.

3. *Authority overlays* – Officers negotiate to plan operations and act cooperatively to maximize output.
4. *Redundancy* – Technical and supply back-ups, decision cross-checks, shadow roles, and the ability to perform more than one task make it possible to replace lost capability.

Four key characteristics are said to typify the high reliability organization (Gaba, 2003):

1. systems, structures and procedures conducive to safety and reliability are in place;
2. intensive training of personnel and teams takes place during routine operations, drills and simulations;
3. safety and reliability are examined prospectively for all the organization's activities, and organizational learning by retrospective analysis of accidents and incidents is aggressively pursued;
4. a culture of safety permeates the organization.

Efforts have been made to export the HRO concept to other sectors beyond the military (Leape et. al., 1998; Bogner, 1994). Significant differences make the transfer of such models from one sector to another difficult, if not problematic. This difference has implications for how we view system performance and system safety. Healthcare provides an opportunity to compare the HRO concept with resilience.

In the case of healthcare, the differences between resilience and reliability are significant.

Healthcare's high variability, diversity, partition between workers and managers, and production pressure make it difficult to employ the redundancy and extensive training that are essential aspects of HROs. This is because healthcare systems do not share the same characteristics as other operational systems such as the naval aircraft carrier. The difference in character makes it difficult, if not futile, to expect healthcare to display results in the same way.

- *Development* – Military systems are developed according to specifications and are maintained according to strict procedural requirements. Healthcare has been likened to a collection of cottage industries that have a loose affiliation (Reid et al., 2005: 12–13).

- *Hierarchy* – Military systems develop and use policy and procedures to improve interchangeability, which enables units and individuals to quickly affiliate and re-affiliate. Career growth relies on candidate evaluation by boards of promotion and screening boards that follow legally mandated guidance. Healthcare follows a commercial and professional model of selection.
- *Management* – Military leaders rise to senior levels of command after training as engineers (generally) and years of exposure to the operational work setting. Healthcare managers are typically businessmen and women who have no clinical background.
- *Behaviour* – The military encourages coordination and, in some services, initiative, based on long-standing service tradition. Healthcare allows for individual caprice if individual practitioners are politically powerful or generate enough billing.
- *Mission* – The military can, and does, stand down to deal with significant safety concerns. Healthcare never stands down.

Among the four key characteristics that Gaba proposed, procedures, training and culture place clinician behaviour at the centre of attention. Given the complex interactions of procedures, equipment, facilities along with personnel, it is not clear whether this is feasible, tenable, or even desirable in healthcare. Very real practical considerations stand in the way of their implementation. Cost limits force systems to run at or near capacity, making the imposition of new systems, structures and procedures problematic. Current pressures to generate revenue allow little room for intensive training beyond what already occurs (Cook and Rasmussen, 2005). Creation of a facility-wide "culture of safety" pales in importance in comparison with the built-in hazards that require engineering, not simply behavioural, attention.

Changes in the type and volume of demand require systems that can adapt, not simply consistently repeat. Demand for healthcare is uncertain, evanescent, and contingent. Organizations to provide care must necessarily change as care demand changes. Healthcare organizations, though, consist of disconnected groups that practice specialties, are only loosely coordinated, and are not well-suited to change. Much of the resilience that has been demonstrated by healthcare systems has been initiated by the human element of the system: clinicians. Wears, Perry, Anders, and Wood (2008), for example, describe how an emergency department staff adapted to handle surges in patient load with no increase in capacity or resources. Physicians and nurses do amount to joint cognitive systems (Hollnagel and Woods, 2005: 21) that can modify their behaviour on the basis of experience in order to maintain order in the face of disruptive influences. However, human operators cannot be expected to provide all of a system's adaptive capacity. Other elements such as an organization's management, equipment, and information technology need to act as a "team player" (Christoffersen and Woods, 2002) for the system to be resilient. Systems that are resilient can sustain required operations even after a major mishap or in the presence of continuous stress. They can mount a robust response to unforeseen, unpredicted, and unexpected demands. They are able to resume normal operations or develop new ways to achieve operational objectives. For these reasons, resilience holds a stronger promise for such a poorly-bounded, uncertain, highly variable, and evanescent work domain (Nemeth et al., 2008).

## Preparation and Restoration

How do organizations prepare to withstand a critical challenge? How do organizations recover from critical events and restore themselves to normal operations? The answers have a great deal to do with resilience engineering and those who seek to understand and use it to improve system performance.

Preparation suggests much more than anticipation or planning. It incorporates all that precedes a challenge: from an organization's structure and ability to adapt and reconfigure, to knowing whether resources can be identified, made available, and defended. Preparation poses questions that are either addressed in these pages or invite further discussion. Have we learned from past lessons, or will the past be repeated in the future? Does an organization have the requisite imagination (Westrum, 1999) that is needed to foresee future challenges? How can various perceptions of the future be integrated to produce a coherent view? Among ideas that are proposed, which of them are worthwhile and how can we tell? Some views of the challenges that organizations face are accurate, but how can they be identified, brought to the attention of others, and integrated? When an organization learns about future challenges, how does the organization respond? Does it encourage the new information? Accept it? Dilute it? Suppress it?

Restoration implies far more than clean-up, recovery, reorganization, or rebuilding. Activities during restoration, following in the wake of an event, can be moulded by a variety of influences. Under pressure to "get on with it," the time to reflect and draw conclusions from what happened is typically shorter than the time that is available to prepare. Unlike preparation, restoration occurs in disorder and in circumstances of degraded performance. The time and effort to restore an organization can vary significantly depending on the amount of damage that was sustained, and the residual effects of recovery can draw out over a substantial length of time. Conclusions regarding causes for the event and what to do about it can be blurred by efforts to shift blame.

Restoration poses similar kinds of questions as we considered for preparation. At what point is it appropriate to start drawing "take away" lessons for the future? What lessons were learned within the organization in the wake of surviving a challenge? How are those lessons viewed? Who is responsible for consolidating the account of what happened, why it happened, and what to do about it? If changes are needed, are they for the short or long term? Are the changes authentic, or simply "lip service" to serve a social agenda? How vulnerable are changes to the tug of "business as usual?" Can changes be made beyond the immediate scope of the organization, or will influences that are outside of the organization push it in directions similar to those beforehand?

Preparation and restoration are not single states that occur sequentially. They are instead ebbs and flows of activity that occur in varying degrees and rates at different parts of organizations and in different neighbouring organizations. Whether either or both occur depends on will, ability, and resources.

**Visible Language of Resilience**

The way that a problem is presented can improve or degrade our ability to solve it (Woods, 1998: 168). Just as tools influence our ability to perform work, our verbal and visual languages influence our ability to cultivate the discipline of resilience engineering. Limits to what our visual language offers may limit our ability in discussions of resilience to express properties such as uncertainty and change through time. This is the visual equivalent of the reductive tendency in thought related to cognitive systems engineering and the substitution myth that it engenders (Feltovich et al., 2004). Many features of tasks such as continuous, simultaneous, and dynamic are challenging to manage cognitively. These and other traits are also difficult for cognitive engineers to allow for in the tools they develop. Our visual representations reflect the difficulty we have with envisioning complexity. As a result, representations show dynamic work as static, interdependent relationships as neatly bounded, and irregular work as regular and routine.

Systems are abstractions and we rely on visual as well as verbal representations to describe them in order to grasp what they could be, what they are, and how they perform. Traditional systems that are developed to operate in simpler and more static circumstances can be represented by simple, static diagrams. However, resilience is substantially about dynamic, not static, properties. New thinking along the lines of resilience requires new kinds of language to describe system properties.

Resilience anticipates future possibilities rather than simply repeating past performance. The future that resilience anticipates is not clear and well defined, but is instead uncertain and poorly defined. The best that our visual language has to offer to indicate uncertainty is dotted, rather than solid, lines. This suggests a binary state between either certain or uncertain, yet a continuum exists between certain and uncertain. The visual grammar does not yet exist to spell out degrees of certainty even though subtle degrees of uncertainty are routine in daily life.

We have similar issues when representing change. Currently, change through time is shown by diagrams such as time lines or the storyboard panels in the manner of a graphic novel or comics.

These can suffice when representing a single dimension, such as time. Resilient systems adapt in various dimensions in addition to time, but we lack the ability to show change in multiple dimensions.

Symbols are easier to recognize when they bear a resemblance to reality (Garrod et al., 2007). Increasingly, though, system elements and types of systems are not part of routine experience. Rather than using symbols, we make gestures as if one can make a vague reference in the hope that the reader will fill in meaning. For example, virtual systems or systems outside of the subject that is of immediate interest are often represented by a meteorological device: a cloud. What means might be developed to represent boundaries or unknown elements that are such a part of systems?

It is possible that animation using electronic media may offer ways to express features such as uncertainty that are essential to discussions of resilience and its engineering. Our ability to create systems that readily adapt also relies on our ability to adapt the verbal and visual media we use to understand and employ the properties of resilience. Nautical charts in previous centuries showed sea monsters at the edge of what was known about the real world. At the moment, our dotted lines to suggest uncertainty and clouds gesturing to distal systems amount to a similar inability to express the nature of what we need to know, but do not; what we need to express, but cannot.

## How the Text is Organized

This book is organized into four sections: Policy and Organization; Models and Measures; Elements and Traits; and Applications and Implications.

*Policy and Organization* explores public policy and organizational aspects of resilience and how they aid or inhibit preparation and restoration. Two chapters move beyond engineering to examine the context that public policy and management creates.

Tom Birkland and Sarah Waterman examine the social construction of resilience: complexity, and incongruity, of public policy. Their analysis shows how conflicting agendas and perceptions across federal, state, and local institutions stand in the way of the consensus that is essential to allocate resources effectively.

Cynthia Lengnick-Hall and Tammy Beck take lessons in resilience from the ways in which commercial firms have dealt with marketplace challenges. While they speak of companies, their insights benefit resilience studies through reflection on the ways that organizations respond in the face of change. Their observations presage Erik Hollnagel's conclusion in the *Elements and Traits* section that management participation will be essential to further the practice of resilience engineering.

*Models and Measures* addresses thoughts on ways to measure resilience and model systems to detect desirable, and undesirable, results.

David Woods, Jason Schenk, and Theodore Allen compare various approaches to models that have evolved in a variety of fields since the 1970s to account for and explain resilience. Their review shows how multiple technical viewpoints shed light on aspects of systems that are related to resilience. It also points to the next challenge that lies ahead: estimating parameters of resilience, so that the concept can be made operational.

John Wreathall describes and contrasts two approaches to system modelling, viable systems (VSM) versus soft systems (SSM), in order to identify traits and processes that they share. Complementing the Woods, Schenk and Allen chapter, he further examines how to develop measurements that relate to resilience.

*Elements and Traits* examines features of systems and how they affect the ability to prepare for and recover from significant challenges.

Erik Hollnagel notes that current ways of thinking about systems extrapolate data from the past into the future in order to estimate risk, but fail to grasp complex emergent interactions among multiple elements. He proposes an approach to thinking about resilience engineering that instead examines properties that systems need in order to be resilient. Effective anticipation of emergent future events must go beyond simple cause-effect relationships. Understanding complex interactions and assuming the burden of risk will require others, including management, to enter the discussion.

Ron Westrum expands on his original conception of requisite imagination (Westrum, 1993). His chapter underscores how insight into system performance in the "human envelope" is available at various levels in an organization, and how tapping such insight can contribute to system resilience.

Philip Smith, Amy Spencer, and Charles Billings focus on restoration using the national air transportation system to illustrate how systems change into a new state after a challenge rather than return to their original condition. They describe multiple approaches than can be taken to imbue resilience: designer anticipation of failure causes, specification that requires response, detection and response to pre-identified scenarios, engagement of novel unanticipated circumstances by relying on the flexibility and intelligence of operators, and the pursuit of more than one approach at the same time as a form of redundancy.

*Applications and Implications* examines how resilience plays out in the living laboratory of real-world operations.

Shawna Perry and Robert Wears present an example at a major medical centre to demonstrate how organizational reliance on safety procedures that exclude human experience and judgment can result in less, not more, resilient performance. They also highlight how the informal network including nurses, managers, Emergency Department (ED) clinicians, and even patients can have an effect on resilience that official means, such as procedures, cannot.

David Mendonça and Yao Hu use an emergency response management exercise to examine how those who are tasked with making decisions handle a severe industrial accident. Their discussion provides an example of mapping real life into the laboratory, creating controlled means to compare responses in different circumstances, then drawing conclusions that suggest further study to improve the understanding of decision-making under pressure.

Martin Nijhof and Sidney Dekker draw on experience in aviation to explore the notion of how to improve the ability of operators to reliably create resilience through improvisation. In the context of two low probability yet high consequence in-flight emergencies, they discuss rules, rule following and abandonment, and the need to develop adaptive capacity by training operators in generic competencies to manage such events.

Emilie Roth, Jordan Multer, and Ronald Scott propose a two-part approach to engineer resilience: anticipation of failure paths, combined with continuously evolvable work practices (developing systems that can be readily modified by users to support their work). The series of "evolvability" features they describe point the way to tailoring software system capabilities to meet changing work needs.

Christopher Nemeth, Michael O'Connor, and Richard Cook consider how IT systems can be developed to contribute to the resilience of joint cognitive systems (Woods and Hollnagel, 2006) in healthcare. Using the infusion device interface as an example, they translate findings from five years of research into an interface design approach that responds to the challenges that Klein et al. (2004) pose for information technology to become a "team player."

## Conclusion

The contents in these pages invite the kind of lively exchange of ideas that is necessary to invigorate any new field. It includes policy and organization studies in addition to engineering and safety science. It addresses the nature of adaptive capacity; how resilience differs from traditional models of system performance; how systems succeed or fail with respect to meeting varying demands; how organizations enable or impede preparation before and restoration after critical events; the trade-offs and costs that enable systems to survive; instances of brittle or resilient systems; the relationship between resilience and safety; and what improves or erodes resilience.

Resilience engineering continues to evolve as a vibrant theme among researchers who deal with system safety, policy, and organizational studies. Insights that each author shares in their chapter point to further opportunities for discussion and research. By the next volume in the series, one hopes that system designers and engineers will also enter into the dialog.

# PART I
# Policy and Organization

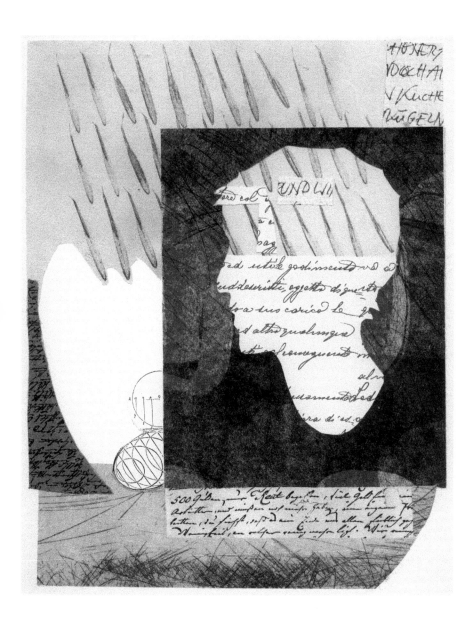

# Chapter 2

# The Politics and Policy Challenges of Disaster Resilience

Thomas A. Birkland
Sarah Waterman

## Introduction

Disaster resilience is an important goal in communities and in engineered systems. The ability of communities, and the systems on which society depends, to "bend without breaking" or to "fail gracefully" has achieved considerable attention among social scientists and engineers, and is being increasingly applied to disaster policy and planning. However, considerable challenges complicate building resilience in social or engineered systems. These challenges include inconsistent definitions of resilience, the lack of political commitment to the broad idea of resilience, the problems of intergovernmental policy formulation and implementation, and problems of citizen perception of resilience efforts. This chapter reviews these challenges both to explain why they are so daunting, and to describe the conditions under which resilience can be achieved. We conclude that efforts to promote resilience already exist in the United States, but that these efforts are unevenly distributed, leaving many communities and their infrastructure vulnerable to long-term damage from which recovery, in the event of a disaster, will be more difficult than it need be.

The apparent increase in the number of natural and technological disasters – and they damage they do – has generated interest in promoting a holistic construct for reducing the damage done as vulnerability increases. Most research and practice suggests that promoting community resilience is a worthwhile endeavour, but moving toward resilience is also remarkably challenging. Defining "resilience" from a societal perspective is challenging because, unlike resilience from an engineering perspective, which measures the ability to absorb a shock, to bend without breaking, or to fail gracefully, features of natural systems are harder to define, identify, and measure. Social and community resilience is therefore metaphoric, but it is indeed a highly useful metaphor. The social construction of resilience – and of the actions taken to make communities more resilient – provides both opportunities and challenges for new ways to look at the idea of resilience and other popular terms such as disaster resistance or sustainability in the hazards and disasters context.

In this chapter we will outline the relationships between resilience and familiar features of hazards and disaster policy and practice. Attention to these aspects of disasters and to the social, political, and economic features of these aspects is crucial to understanding prospects of resilience as an organizing principle and as a social or policy goal.

## Engineering and Social Perspectives on Resilience

While there are many ways to define disaster resilience (Comfort 1994; Aguirre 2006; Manyena 2006), most definitions of resilience encompass the idea of "failing gracefully" or having "rebound capacity." Definitions of "resilience" derive from various fields, but two of particular note are ecology and engineering. In ecology, a resilient ecosystem is one that can absorb some sort of a shock to the ecosystem and then return, ultimately, to providing the same level of ecosystem services that were provided before (Adger et al., 2005). For example, a small oil spill in an estuary may lead to short-term environmental damage, but, over time, as the volume of remaining oil in the ecosystem declines, and as the system recovers from the shock, the estuary returns to its same level of function. Some ecosystems, however, fail to return to their prior level of functionality, either because the initial damage is so profound, or because the damage is just one of a series of incidents that overwhelm the ecosystem's ability to rebound and respond (Wheelwright 1994; Wiens 1996).

The engineering idea of "resilience" involves engineered systems that can be stressed to a particular point without breaking, or without suffering a significant degradation in the service the engineered system provides. For example, in an earthquake, a *resilient* bridge may lose its ability to bear normal loads, say from trucks, but remains intact so that it can be repaired, rather than having to be entirely or nearly entirely rebuilt. Bridges that collapse in earthquakes are by definition not resilient at a particular level of stress. Such bridges remain standing until catastrophic failure results. The goal in resilient engineered systems is either to allow for a wide performance envelope given particular stresses, or to allow some components of a system to fail so that the overall engineered system does not fail. In the former example, an aircraft wing is designed to be highly tolerant of a wide range of loads beyond those expected in normal operations. A bridge that flexes to the point of substantial deck failure, but which preserves its overall structural form, can be said to have failed gracefully. We can think of communities in the same vein. A community needs to be able to bear stresses without losing all functionality if it is to be able to recover from a natural disaster.

A similar idea is reflected in the systems literature, where resilience is offered as a solution to the open system nature of increasingly complex human systems (Fiksel 2003). Systems scholars argue that since anticipating every possible shift in the environment is impossible, complex systems require a high level of built-in resilience. Fisk (2004) cautions that complex systems, however, may not manifest built-in resilience. This is because so much time passes between disasters. When these external shocks disrupt the system, the consequences can be dire. New Orleans' experience of Hurricane Katrina provides an example of these disguised vulnerabilities, although, to be fair, some were less latent than others. Leading up to 2005, the city had cautiously embraced the idea of "near misses," ignoring obvious vulnerabilities because the previous hurricane seasons had been remarkably kind. Furthermore, expectations that the federal government would provide sufficient aid gave an illusion of invulnerability. Even as Hurricane Katrina bore down, assurances of the comprehensive response plan masked problems (Cooper and Block 2006). From a systems perspective, the failures experienced during and after Hurricane Katrina can be partially explained by the significantly complex political, economic, and social systems operating during the storm. And, as Perrow (1999) notes, many complex systems create unintended consequences as a result of unanticipated interactions between systems that, taken individually, are supposed to enhance safety rather than create unanticipated and therefore hard to solve problems.

It is noteworthy that the *political* considerations or features of the policy process in local, state or regional politics rarely feature in analyses of resilience (Kendra and Wachtendorf, 2006). Yet, the nature and function of politics is as important to community resilience as any other social process. Politics is an activity that helps societies decide "who gets what, when, and how" (Lasswell, 1958). Disasters and the policies that deal with disasters are clearly political in the sense that they imply the distribution of resources through a system of government. Thus, we may offer the hypothesis that most if not all aspects of resilience are influenced by public policy, either by design or by accident.

Researchers and practitioners have recently begun to be more interested in resilience and vulnerability. This interest may be driven by catastrophic events such as Hurricane Katrina and the 2004 South Asia earthquake and tsunami, which highlighted a form of resilience in which a community can begin to recover and provide aid to victims for an extended time without outside aid (Manyena, 2006). This interest may also be driven by a desire to find "all hazards" terminology to address natural, technological, and purposive hazards in an era of extreme attention to terrorism and reduced concern about natural disasters, a trend that Hurricane Katrina may have reversed. The emerging discussion of resilience is helpful because it might be a way to think more holistically about the disaster cycle: instead of thinking of preparedness, mitigation, response, and recovery as different phases of a cycle, we can think of them as aspects of a system that promotes or retards resilience.

An argument for policies promoting resilience is that only relatively marginal improvements in communities are needed to be successful; although there are many such improvements that need to be made. As Campanella (2006) notes, most cities, in particular, already possess a strong tendency toward resilience.

Even when badly damaged, cities contain the fundamentals: a group of surviving citizens, and the core of a city that exists on that site for some reason. Even when infrastructure systems are not resilient, the overall city itself has considerable resilience.

The role of engineering in the public policy process is sometimes misunderstood by engineers and non-technical policy-makers. Engineers are motivated by many goals, including the physical strength and resilience of systems, their operational efficiency, efficiency in their construction, their safety (a paramount concern) and, of course, the elegance of the system or a component of a system, ranging from the technical excellence of a telecommunications system to the beauty of bridges and similar public works. Ultimately, engineering is both a creature and servant of society, and, as such, is embedded in all the other social aspects of communities that promote or inhibit resilience (Petroski, 1992). Engineers, however, are uniquely positioned to understand the nature of infrastructure, the interrelatedness of infrastructures, the degree to which infrastructure systems can absorb strain or shocks, and the extent to which the system has designed or serendipitous resilience. And from engineering comes the important metaphors that define communities: robustness and resilience. Engineers play a crucial role in promoting community resilience; they must also consider the social and political environment of engineering, as well as the limits of engineering in protecting communities from the worst outcomes of disasters. The subtle balancing of social, political, economic, environmental, and engineering aspects of communities is particularly difficult to achieve, but is a worthy goal in a world of increased hazard vulnerability. This chapter will help outline some of the features of the environment in which engineers and all decision-makers must operate.

## Resilience and Vulnerability

Researchers at MCEER (formerly the Multidisciplinary Center for Earthquake Engineering Research) at SUNY Buffalo have developed the idea of a "resilience delta" (MCEER, 2006) that reflects three dimensions of resilience: pre-disaster functionality of an infrastructure system, the extent of damage to the infrastructure system, and the speed of recovery of that system. Their vision of resilience is reflected in Figure 2.1.

Their "R4" notion of resilience includes:[1]

- Robustness – strength, or the ability of elements, systems, and other units of analysis to withstand a given level of stress or demand without suffering degradation or loss of function;
- Redundancy – the extent to which elements, systems, or other units of analysis exist that are substitutable, i.e., capable of satisfying functional requirements in the event of disruption, degradation, or loss of function;
- Resourcefulness – the capacity to identify problems, establish priorities, and mobilize resources when conditions exist that threaten to disrupt some element, system, or other unit of analysis (resourcefulness can be further conceptualized as consisting of the ability to supply material – i.e., monetary, physical, technological, and informational – and human resources to meet established priorities and achieve goals);

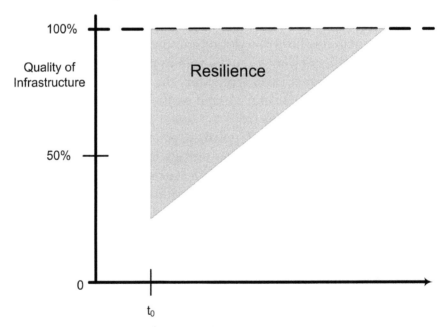

**Figure 2.1    The resilience delta**

---

1          The language in the lists that follow is directly quoted from MCEER, 2006.

- Rapidity – the capacity to meet priorities and achieve goals in a timely manner in order to contain losses and avoid future disruption.

These features of a resilient system are contained in what MCEER calls the four dimensions of resilience:

- Technical – the ability of physical systems (including all interconnected components) to perform to acceptable/desired levels when subject to disaster;
- Organizational – the capacity of organizations – especially those managing critical facilities and disaster-related functions – to make decisions and take actions that contribute to resilience;
- Social – consisting of measures specifically designed to lessen the extent to which disaster-stricken communities and governmental jurisdictions suffer negative consequences due to loss of critical services due to disaster; and
- Economic – the capacity to reduce both direct and indirect economic losses resulting from disasters.

In a similar vein, Aguirre notes that communities have different resilience profiles based on "physical, biological, psychological, social, and cultural systems" (Aguirre, 2006: 1). Similar thinking has informed research and programs on "disaster resistant" communities, although it is important to note that 100 percent *resistance* to a disaster is not usually achievable, from economic, political, or engineering perspectives. Rather, a resilient community is one that may not be entirely resistant – that is, things may break during the disaster – but that has the capacity to recover quickly, both because vulnerability is reduced and resilience is promoted. If we conceive of a community faced with many events – hurricanes or coastal storms, for example – a learning and resilient community would be one that experiences a series of disasters that become less disruptive, presumably due to learning and the application of experience (Birkland, 2006).

The disaster delta is an important concept because it contains three key aspects of resilience: the extent of initial damage to a system, the rate at which recovery occurs, and the level of functionality to which the community returns. While MCEER is primarily concerned with infrastructure resilience, it does not require a great leap of imagination to view the vertical dimension as overall damage to the community. This is not only measured monetarily, but is also measured in terms of functionality of the community. Functionality includes the extent to which the community works as a community, with people, organizations, infrastructure, and government supporting the activities that characterized the community before the disaster. In Figure 2.1, the horizontal line along the top extends from the 100 percent functionality of the community. The hypotenuse of the delta simply shows the rate of recovery to the prior level of functionality.

In Figure 2.2, the basic model is altered in some substantial ways:

- The line representing Community 1 is the basis for comparison.
- Community 2 is less resilient because the length of time it takes to recover from the same shock as Community 1 is considerably greater.
- Community 3 is also less resilient, because, although rapidity and resourcefulness are great, the community is less resilient

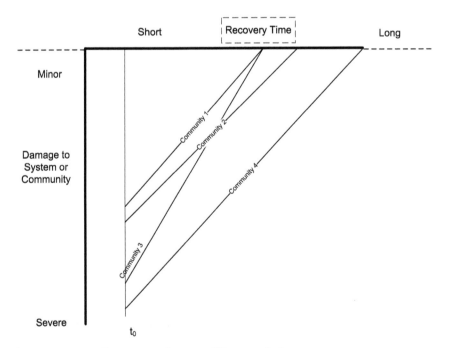

**Figure 2.2**     **Comparative resilience deltas**

overall because initial damage is so great. In other words, the costs and challenges of resilience are represented in all these cases by the area of the delta, not merely the depth of damage or the extent of recovery time.

- Community 4 is the least resilient community because, even though the slope of the recovery line is the same as that in community 1, damage is more severe, and recovery time takes longer.

In Figure 2.3, we see two examples of other outcomes of disasters in truly nonresilient communities. In both lines, the extent of the damage is relatively high, but we also assume that the community never returns to pre-disaster levels of functionality. The rebound lines represent the extent to which the community has "rebounded" but only to pre-disaster levels of functionality, and the extended line therefore helps describe the area above the functionality line that represents the long-term losses to the community or the system.

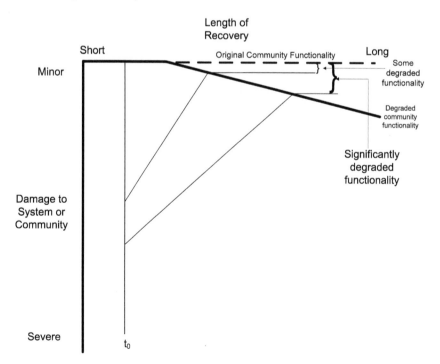

**Figure 2.3**     Resilience deltas when community function is permanently degraded

When viewed this way, it is clear that there are three key features of resilience: the prevention of damage, the speed of recovery, and the prevention of substantial decay in the functionality of the community. Community functionality is not the same as a return to the *status quo ante*, which often replicates the very vulnerabilities that lead to poor resilience in the first instance. Rather, the community would, it is hoped, return to its previous level of social and economic function while reducing vulnerability. These goals are often viewed as incompatible, particularly if people express a strong desire to return to "normalcy," if normalcy also reproduces pre-existing vulnerabilities.

These resilience deltas are highly simplified, of course. Communities are composed of many systems – infrastructure systems, healthcare, schools, private industry, and the like, each with their own resilience deltas that, when taken together, provide a fuller picture of community resilience, particularly when we note that these systems are tightly coupled and highly interdependent.

## The Relationship between Resilience and Vulnerability

There is considerable evidence of increasing vulnerability to disasters in the United States and worldwide (Green et al., 2006; Cutter and Emrich, 2006; Lewis and Mioch, 2005; Jacobs, 2005; Mustafa, 2003). The trends that increase vulnerability will likely continue in the near future, even in the United States. While it might be tempting to believe that a wealthy and advanced nation has few places vulnerable to major disasters (or catastrophes), recent experience suggests that areas of the nation are vulnerable to catastrophic disasters. The increased number and costs of natural disasters results from population growth in vulnerable areas. For example, coastal areas in the United States are particularly prone to hurricanes, earthquakes, tsunamis, floods, fires, and landslides. Often these hazards run together, such as the combined landslide, fire and earthquake hazards in southern California (Davis, 1998). Better responses will be a function of better state and local preparedness and mitigation efforts, such as states with substantial disaster experience. California, Florida, and North Carolina have learned from disasters and seek to improve performance based on that knowledge (Birkland, 2006; Geschwind, 2001; Mittler, 1997; North Carolina Division of Emergency Management, 2004; North Carolina Office of the Governor and North Carolina Office of State Budget and Management, 2006; Work et al., 1999). Even then, we cannot claim that better prepared and more resilient states and communities have reduced vulnerability among the poor, women, children, and the elderly to the same extent as the broader population (Cutter and Emrich, 2006).

In other states, disasters and catastrophes will exist and may worsen because few attempts have been made by the federal government to encourage states and localities to assess their vulnerabilities, assess hazards, and take disaster mitigation steps; indeed, even homeland security funding is not strongly risk based (Reese, 2005), but, instead, is allocated based on redistributive politics ("pork"). With no incentives to mitigate (and plenty not to, such as the relentless pressure to develop real estate [Burby, 2006]), many communities continue to increase their vulnerability. The federal executive branch will continue to use copious amounts of disaster relief as a political and economic palliative (Platt, 1999), while states and local governments follow local political and short-term economic incentives to rebuild as quickly as possible in the same vulnerable ways and places (Alesch and Petak, 2001, 1986). Economic activity and tax revenue return to the *status quo ante*, but vulnerability exists or even increases. Indeed, outside aid may well create a "moral hazard," in which people and communities are unmotivated to mitigate and to promote resilience because they think that the federal government will be the insurer and responder of last resort (Cutter and Emrich, 2006; Allen, 1997).

This scenario describes the situation on the Mississippi Gulf Coast after Hurricane Katrina. In Biloxi, for example, disaster victims have sold their damaged properties (sometimes for handsome sums) to growing, larger Gulf Coast casinos or high-rise luxury condominium developers. While these new buildings are somewhat more robust than the ones they replaced, vulnerability will increase as higher population densities are placed near hurricane zones; at the same time, tourists, a challenging population from a disaster management perspective, must be encouraged to be vigilant and to evacuate in a timely manner.

An alternative way of examining vulnerability suggests that vulnerability is the eye of the beholder, and that our primary concern should be with resilience. In this conception, Furedi (2007) argues that researchers often assume that a community is vulnerable, with little consideration given to the inherent resilience of most communities. Thus, vulnerability is revealed when one adopts a particular epistemological framework in which vulnerability is considered a more common feature of communities than is resilience, and "since [researchers] believe that 'societies and communities are always vulnerable,' it is seen to define the condition of existence." (Furedi, 2007: 487).

If we adopt the vulnerable community model, we also adopt with it, at least to some extent, the helpless community model, with its concomitant needs for rapid outside assistance of all types. Because of the political attractiveness of generous disaster relief, this vulnerability perspective has corresponded with the "defining downward" of a "disaster." In the 1990s, the Clinton administration, realizing the political benefits of relief, defined many previously unremarkable events as disasters, such as large snowstorms in northern climates (Allen, 1997; Platt, 1999). Such declarations allowed federal money to help pay for snow removal, a task that was generally budgeted for and paid for by local communities. Even relatively small natural events became disasters, such as the tornado that struck communities near Albany, New York, in 1998, where aid was made available through the Federal Emergency Management Agency's public assistance program, which provides funds to local governments to repair infrastructure and defray response costs. Some funds were also made available to property owners through the US Department of Housing and Urban Development even though most people were covered by insurance and the total damage was $50 million, or about 0.001 percent of New York State revenues in 1998. There are clear political reasons for this defining downward of disasters, relating to the "deservedness" of the aid and the claimed virtue of the victims. Some of these victims unwittingly contribute to their own vulnerability, although others are vulnerable through no fault of their own.

Thinking about resilience and about vulnerability together, although not as mirror images of each other, is important. In particular, we can view vulnerability as something that lengthens or shortens the vertical line on the recovery delta (Figure 2.1), with vulnerability being the conceptual rubric for the things that make a community's initial disaster greater or lesser. The angle between the vertical line and the hypotenuse represents both vulnerability and, in particular, resilience, because the angle (in this very simplified depiction) is what determines the time to full recovery. This causes us to think slightly differently about the four Rs of resilience, with robustness and redundancy being features that would reduce damage along the vertical axis, while resourcefulness and rapidity are features of systemic or community resilience and vulnerability that are reflected on the horizontal axis. Thus, robustness and redundancy can maintain community functionality at a higher level than a non-robust, non-redundant community, but recovery time – resilience – can vary considerably even when a system is robust. Compare, for example, communities 1 and 2 in Figure 2.2 – both are equally robust, but community 2 is less resilient. The nature of this rebound is a function of both resilience and of vulnerability. In other words, resilience and vulnerability are complementary, but not congruent.

Of course, the meaning of "robustness" and "redundancy" is subject to considerable interpretation and definition, and precise definitions are beyond the scope of this chapter. For current purposes, we can say that community robustness reflects the community's ability to absorb an event with relatively little disruption. For example, we can say that New York City was generally very robust after the September 11 attacks and the failure of the World Trade Center (WTC), because the city of New York absorbed the shock of September 11 rather well. While it is undeniable that the attacks were spectacular and highly disruptive, the city of New York is large enough and has sufficient resources to stand up to the attack, respond to it, and clean up after it (Harrald, 2006). While outside help was made available, much of New York's response was in house, from emergency response to the management of debris removal (Langewiesche, 2002). The loss of over 350 firefighters and over 50 police officers was truly a major blow. Yet, the city's emergency services were large enough and robust enough to be able to respond, two months later, to the crash of American Airlines flight 587 in Queens on 12 November 2001 (Wakin, 2001), and to attend to routine matters.

We can also say that New York had substantial resourcefulness, in two dimensions. The city had actual resources – personnel, expertise, equipment, and the like – to respond to the disaster. Even when 7 WTC collapsed, taking with it the Emergency Operations Center (EOC), the city was ultimately able to use its resources to establish a makeshift EOC on Pier 92 on the West Side of Manhattan (Kendra and Wachtendorf, 2003; Dawes et al., 2004). Nevertheless, the establishment of such a centre would not have been possible without the resourcefulness of city and community staff. This resourcefulness aided the speed with which the city was able to respond and begin to recover from the event. One week after the disaster, the stock market reopened, and downtown businesses started the recovery period that continues today, as the new WTC complex is built.

Thus, focusing on resilience starts with a set of assumptions that are well documented in the literature on disasters, going back at least to the Second World War. We know that communities are generally resilient, that people can form emergent prosocial groups, that people do not riot, loot, or panic, and that even badly damaged communities contain within them considerable potential for recovery (Dynes, 2003). While recognizing that different populations may be more vulnerable than others, we can refocus the discussion on how to induce resilience in two ways: overall resilience in the community, and the promotion of greater resilience among the most vulnerable populations. We can thus stipulate that very few communities are so fragile that they are unable to recover in some way (Dynes, 2003). A key question, of course, is to what extent communities recover to pre-disaster functionality, particularly without reproducing the same features of the community that created or amplified vulnerability in the first place.

## The Disaster Cycle and Resilience

The disaster cycle is a common feature of all treatments of disaster and hazards. It is a cycle because there is no end: nearly all communities in disaster-prone areas will experience a disaster, will respond to it, recover from it, and prepare for it. Many, but not all, of these communities will attempt to mitigate the damage done by disasters. In the American experience, communities are more likely to mitigate disasters through engineered systems – levees, dams, stronger buildings – than through changes in individual or collective behaviour.

Preparedness is a key feature of resilience that, at a given level of damage, can lead to actions that prevent damage from worsening, and can hasten recovery. While a community that does not engage in effective mitigation may still have a very deep hole to dig out of, preparedness is one of the aspects of emergency management that will shorten the period between the acute phase of the event and a return to something approaching normal. This chapter will consider mitigation more carefully, but we acknowledge that disaster preparedness also influences resilience. Preparedness and mitigation therefore overlap because they both seek to reduce long-term damage to a community. But mitigation is more about reducing the extent of damage, while preparedness is about responding to any given level of damage – the angle in the resilience delta described above.

For this chapter we use FEMA's definition of mitigation: "Mitigation is the effort to reduce loss of life and property by lessening the impact of disasters." FEMA continues to define mitigation in rather narrow terms: "[Mitigation] is achieved through risk analysis, which results in information that provides a foundation for mitigation activities that reduce risk, and flood insurance that protects financial investment" (FEMA, 2005).[2] At the other end of the spectrum, Mileti (1999) argues that mitigation is central to community sustainability, in which communities work to maintain and enhance environmental quality, improve overall quality of life, create resilience, and assume local responsibility for hazard mitigation and, therefore, resilience. If this path were followed, equity and overall quality of life would result, and major differences in vulnerability would erode. This is an expansive and, given the continuing uncertainty over what the term "sustainable" really means, a politically unrealistic position

---

2       I thank Sean Hildebrand for pointing out these two differing definitions.

in the short term, but Mileti's vision does suggest the blending of community development, sustainability and mitigation in a way that ultimately enhances resilience.

The classic rationale for pursuing mitigation is simple: it prevents future losses, it therefore puts less stress on response systems, and simultaneously leaves a community better prepared to recover (Multihazard Mitigation Council, 2005). Mitigation is typically divided into two categories: structural mitigation and nonstructural mitigation. The Multihazard Mitigation Council (2005) divides mitigation measures into "project" and "process" mitigation. Project mitigation involves building things that resist natural forces, such as levees, beach groins, beach nourishment, floodwalls, and the like. Process mitigation involves using policy tools to alter the behaviour of actors in the process, so that their actions will promote or at least not undermine community, regional, and national mitigation goals. Process mitigation includes planning, zoning, building codes, and similar regulations intended to reduce the vulnerability of a structure, site, or neighbourhood to disasters, thereby enhancing overall community safety.

As suggested by Mileti, the goal of any disaster mitigation program should be community resilience (see also Godschalk, 2003), but promoting mitigation is not the same thing as promoting resilience, particularly when project mitigation, in particular, often undermines long-run resilience even when it prevents short-term nuisances (Burby, 2006). For example, some project-oriented mitigation works, such as levees, might prevent most flooding but may make flooding worse during a catastrophic failure. Like the idea of masked vulnerability from the systems literature, such works therefore replace resilience with robustness, but robustness implies strength until a catastrophic failure occurs (the opposite of "failing gracefully") thereby yielding catastrophic damage in areas where people believed that the mitigation works would protect them. This is generally considered a "false sense of security," particularly when the claimed or perceived level of robustness falls short of actual performance, as happened in New Orleans during Hurricane Katrina (van Heerden, 2007; Seed et al., 2006; Kintisch, 2005). Resilience is compromised because the levee encouraged people to build more intensively in the hazardous area than would have occurred had a levee not been there (Birkland et al., 2003). This problem extends to federally subsidized insurance and, potentially, disaster relief. When people believe that someone will help them recover from risky decision-making, they are less prone to reduce risks, and are more prone to take risks – this is called a "moral hazard" problem.

Process mitigation, such as land-use planning to remove vulnerable property from hazardous areas, may increase resilience simply because mitigation reduces overall damage – the sharp downward line in the resilience delta. Furthermore, process-oriented mitigation programs may engage people and community institutions more directly than project mitigation will, thereby creating knowledgeable and supportive communities that understand and seek hazard mitigation (Burby et al., 1999; Burby and May, 1998) and that are prepared to mobilize after disasters. Such mobilization is a key feature of post-disaster resilience.

Little changed in the 1990s to reduce the subsidies of risk that Burby cites. However, some efforts were made by the Clinton administration, and by FEMA under James Lee Witt, to support local mitigation initiatives, such as through Project Impact, an effort to engage local governments and the private sector in disaster preparedness and mitigation (Murray, 2001; Freitag, 2001). Local hazard mitigation became an important element of disaster policy in the 1988 Stafford Act and in the 1993 Stafford Act amendments. The shortcomings of the 1993 amendments – particularly with respect to harmony between state and local mitigation plans – were addressed in the Disaster Mitigation Act of 2000. Project-related mitigation remains important both for substantive mitigation reasons and because such projects, for example flood control works undertaken by the US Army Corps of Engineers, are broadly popular among members of Congress who actively seek such distributive spending, even as such projects subsidize risky behaviours and do not work particularly well over the long run. These techniques also carry with them substantial environmental side effects that likely diminish whatever the net value of the project is to society (Birkland et al., 2003). The efficacy of many of these projects is less important to many members of Congress than is their distributive nature. These projects therefore allow for credit claiming and the distribution of "pork barrel" projects to individual districts (Murray, 2001; *Houston Chronicle*, 1995). While some claim that these projects are simply handouts to those unwilling to reduce the negative consequences of their own risky actions, they are particularly difficult to resist because of their political benefits (Prater and Lindell, 2000).

## Why Current Policies Inhibit Resilience

The most popular and most common federal expenditures relating to natural disasters are for relief and recovery. Funding is provided to communities through FEMA's public assistance program, and direct grants of federal aid to individuals are available from FEMA's individual assistance program in declared disaster areas. The Small Business Administration (SBA) offers disaster loan programs for individuals and businesses. Such payments are not wildly munificent – FEMA makes clear in its guide for disaster victims that Individual and Household Relief (IHR) under the individual assistance program is not intended to supplant insurance, will not cover all losses, and is often secondary to SBA loan programs (FEMA, 2005). FEMA provides aid to communities through the public assistance program, which helps fund debris removal, infrastructure repair, and similar community costs. Taken together, FEMA disaster relief seeks to compensate disaster victims for what Burby (2006) calls "residual risk" – that is, the risks that remain after flood protection, hurricane protection, beach nourishment, or other project-oriented mitigation actions are taken.

Of course, such policies yield substantial moral hazard problems, and often simply don't work. Beach nourishment is a Sisyphean task that involves constant restoration of beaches in the face of continuing demand to restore beaches for the benefit of property owners. This pressure continues even as that very development disrupts normal sand transport mechanisms (Kaufman and Pilkey, 1983; Dean, 1996; Pilkey and Dixon, 1998). Building elevation and flood insurance requirements often fail to take into account changes in the flood plain resulting from additional development in the floodplains; these changes include both the extent of the floodplain and the depth of floodwaters. Moreover, when engineered flood and hurricane protection works fail, they often do so catastrophically. For example, the 100-year flood – that is, a flood with a one per cent probability of occurrence in any given year – might not be the best design standard for such works. Given the possibility of catastrophe, probabilistic thinking should be considered as seriously as "possibilistic" thinking – "what's the worst that could happen?" – in areas where catastrophic failures would be catastrophic for the region (Clarke, 2005a, 2005b). Even then, not every hazard-prone community can be solely protected by constructed mitigation, because we cannot know what the worst outcome is, and we often cannot afford to build for it. And some hazards, such as earthquakes, cannot be mitigated through the building of major public works projects; rather, mitigation is nearly always process-oriented, and is focused primarily on building codes and on improvements in professional practice (Geschwind, 2001).

The shortcomings of federal disaster policy and their distortions of state and local regulation, and of individual behaviour, are well known, and these issues are repeated after every event. If this is the case, and federal effort could be shifted away from relief and insurance payouts (that is, promoters of so-called "moral hazard") toward preparedness and mitigation, it is likely that the rate of growth of disaster losses, if not the absolute value of disaster losses, could be reduced.

## Promoting Resilience in Public Policy

One of the most difficult aspects of policy design is the choice of the "tools" that government will use to induce desired behaviours (Salamon and Lund, 1989). These behaviours are induced in the target populations, be it the governments, individuals, or nongovernmental organizations such as businesses or charities. There are several important features to keep in mind in national disaster policy over the past 60 years.

First, national policy in the United States (and in other federal systems), due to constitutional constraints and political realities, does not and cannot mandate that particular mitigation or preparedness actions take place (Burby, et al., 1996; May, 1993, 1994, 1995; May and Birkland, 1994). Rather, national policy can simply provide an *inducement* to state and local governments to take steps that would promote resilience. But there are few, if any, guarantees in policy-making that any particular policy will be implemented unless there is broad agreement among all parties that the policy should be implemented in a particular way (Goggin et al., 1990).

Second, federal disaster policy is generally piecemeal, does not consider disaster resilience as a major organizing principle, and, in fact, many, if not most, federal policies fail to work together to promote resilience. For example, structural mitigation projects such as floodwalls and levees can reduce nuisance flooding, but when these structures fail catastrophically, the resulting flooding often does substantial damage.

Third, disaster policy at all levels of government is remarkably political, in the sense that it represents some sort of vision of the allocation of resources. As Rutherford Platt notes (1999), starting in the 1990s, the decisions on whether to grant disaster relief were apparently driven as much by political concerns as by some objective sense of "need" or "deservedness." Rather, the Clinton administration, learning from the debacle that was the federal response to Hurricane Andrew, began much more generous grants of disaster relief to communities struck by a "disaster," often with much less regard for whether and to what extent the community had the capacity to address a disaster on its own. Communities were of course entirely willing to accept such generous aid, and were in fact eager to do so. An attempt was made with the 1993 Stafford Act Amendments and with the Disaster Management Act of 2000 to encourage greater community mitigation planning as a requirement for receiving mitigation funds. However, mitigation funding required that a disaster take place first. Only then would the federal government provide funds under the Hazard Mitigation Grant Program (HMGP). However, at its peak funding level such funds represented only 15 percent of the funds spent on disaster relief. Other mitigation efforts, such as Project Impact, have been cut by the Bush administration, to the point where federal support for disaster mitigation is even less evident and less effective than it was before Bush took office. At this writing, it is unclear whether mitigation will become a priority under the Obama administration.

Fourth, state and local policies – and their implementation – are functions of prevailing local conditions, community wealth, experience with disasters, and many other factors not amenable to a single set of policy tools or solutions. Communities prepare for disasters with significantly different resources, objective risks, risk perceptions, and considerably different hazard profiles. Because of these differences, federal policy influences state and local disaster policy, but is not sufficiently prescriptive or coercive as to enforce resilience-promoting policies.

The differences between federal, state, and local goals are clear if we look at emergency management practice after the September 11 attacks. The state and local emergency managers, even after September 11, took the "all hazards" approach much more seriously than the federal government did. The federal government is currently much more concerned about terrorism, while state and local governments have consistently realized that they are much more likely to have to address a range of natural and technological hazards and disasters as well as prepare for the much less likely eventuality of terrorism. The federal government's focus on homeland security therefore is not antithetical to state and local goals, but it is not currently congruent with state goals, either. This stands in rather sharp contrast to the situation during the 1990s, when there was considerably more cooperation between the federal state and local governments. One can therefore say that federal policy, on the one hand, and state and local policy on the other have been divergent since 2001 (Scavo et al., 2008).

## The Political Feasibility of Resilience in a Federal System

Beyond the question of response, however, one must consider the federal role in inducing state and local governments to allow development in hazardous areas, and the intergovernmental role in persuading people that they were probably safer than they really were. Ray Burby (2006) calls these the "safe development" and the "local government" paradoxes. In the former, Burby argues that the federal government, through all manner of subsidies and incentives, including flood insurance, levees, tax deductions for casualty losses, and generous disaster relief, encourages development in areas that would not otherwise be developed if the risk profile of the area were undistorted by these incentives. People whose risk perceptions are altered in this way are said to be susceptible to "moral hazard," in which people take risks they would not ordinarily assume because they believe that the additional risk is attenuated by the protection provided by government. Of course, these decisions may well be distorted by perceptions: building behind a levee may provide the illusion of safety, but the probability of the levee failing because of poor design, poor construction, poor maintenance, or of exceeding the design flood are not often considered in the risk calculation. One can certainly argue that vulnerable areas are "safer" from routine events, such as "nuisance flooding," but whether an area is truly "safe" is a matter of probability just like any other risk.

The local government paradox suggests that even when the federal government wants to address hazards more aggressively, the state and local governments must implement federal ideas for mitigation. Students of intergovernmental relations and policy implementation know that designing such policies is very difficult. Malcolm Goggin and his colleagues argued in 1990 that intergovernmental policy implementation is more likely to be successful when credible policy designers communicate clear and credible signals to capable implementers whose interests are congruent with those of the policy originator (Goggin, 1990). Unfortunately, the interests of the federal government are not clear, are inconsistent, and are not clearly communicated. Even when an incentive, such as the Hazard Mitigation Grant Program, is provided, it may not be taken up by the local governments, or their ideas of what constitutes sound mitigation may not be congruent with those of federal policymakers.

At the same time, the paradox of local development is also at work. As Rutherford Platt notes, the costs of nearly all effective preparedness and mitigation measures are borne by state and especially local governments in at least two ways: in the funds that have to be expended to execute these plans, and in the loss of tax revenue and other economic activity that accompanies land that would otherwise be developed. The promise of increased taxes and funds was what induced the Orleans Parish levee board to use levees more as a way of creating new developable land than as a hazard mitigation measure (Cooper and Block, 2006).

These actions, by their very nature, increased vulnerability. This problem is not unique to New Orleans. Many communities, particularly those running out of developable land, experience the same pressures. Local property tax revenues are the major source of municipal revenue, and urban "growth machines" (Logan and Molotch, 1988) work with civic leaders to promote development in unsafe areas (May, 1995; Burby et al., 1996, 1999; Burby, 1998a and b, 2006).

## Prospects for Resilience

This chapter has reviewed aspects of resilience and the political and policy factors that impede resilience.

We know from historical examples that communities can fail to fully recover from disasters because of political, economic, or social reasons that relate to the disaster combined with historical factors that coexist with the disaster. Galveston, Texas, never really recovered after the 1900 hurricane because of economic competition from Houston, fifty miles inland. And New Orleans may never return to its size and economic importance, not just because of the hurricane, but because of loss of business to Houston and other regional hubs as the oil industry consolidates and as ports become more competitive. Disasters can therefore be catalytic events, but are simply part of the history and fabric of any region's broader social, economic, and political history. It is up to all stakeholders, and their expert advisors, to develop and implement a vision of how communities can recover and remain resilient, with a particular focus on the social, economic, and historic environment in which a community is located.

Pessimism about a community's post-disaster fate should be tempered by a plain fact: some level of resilience is inherent in every community. Resilience comes from the place the community occupies spatially, but it also comes from the sociocultural place in the community, and the economic system in which it is embedded. Thus, throughout history, cities that seemed badly wounded did bounce back, sometimes better than before. After all, the Great Chicago Fire of 1871 was certainly catastrophic, but it also impelled the city to alter fire codes, change urban planning standards, and, ultimately, to develop an entire "school" of architecture, with the aim of becoming the dominant commercial and cultural hub in the Midwest. Anchorage, Alaska, rebounded quickly after the 1964 earthquake because of outside aid, the importance of the city in the state's economy, and, ultimately, because of the explosive economic growth in the state that followed the discovery of oil on the North Slope in 1968. New Orleans will also recover, eventually, from Katrina: it is an important port, it serves the off-shore oil industry in Louisiana, and, of course, its cultural importance looms large.

With these devastating examples in mind, scholars and practitioners should take up the challenge of understanding how the scientific, technical, natural, and social realms all influence community resilience and recovery. We know it can be done, because it has been done before.

# Chapter 3

# Resilience Capacity and Strategic Agility: Prerequisites for Thriving in a Dynamic Environment

Cynthia A. Lengnick-Hall
Tammy E. Beck

An organization's resilience capacity captures its ability to take situation-specific, robust, and transformative actions when confronted with unexpected and powerful events that have the potential to jeopardize an organization's long-term survival. Strategic agility is a complex, varied construct that can take multiple forms but captures an organization's ability to develop and quickly apply flexible, nimble, and dynamic capabilities. These organizational attributes share common roots and are built from complementary resources, skills, and competencies. Together, strategic agility and resilience capacity enable firms to prepare for changing conditions, restore their vitality after traumatic jolts, and become even more proficient as a result of the experience. Resilience capacity helps firms navigate among different forms of strategic agility and respond effectively to changing conditions. In this chapter, we explain why organizational resilience capacity can be viewed as an antecedent to strategic agility, and as a moderator of the relationship between a firm's dynamic activities and subsequent performance.

Every Scout has learned the motto "Be Prepared." For organizations, being prepared means that a firm or agency is equipped to deal with unforeseen adversity and it is ready to capitalize on unexpected opportunities. In turbulent, surprising, continuously evolving marketplace environments only well-prepared, flexible, agile, and relentlessly dynamic organizations will thrive. Unstable environments create frequent challenges. Often these events are viewed negatively, but as Sutcliffe and Vogus (2003) explain, resilient organizations are able to maintain positive adjustments under disruptive conditions. Resilience capacity provides the basis for restoration after a severe jolt and can offer an opportunity for an organization to undergo a positive transformation as a result of overcoming an exceptionally challenging experience. Similarly, strategic agility enables a firm to initiate and apply flexible, nimble, and dynamic competitive moves in order to respond positively to changes imposed by others and to initiate shifts in strategy to create new marketplace realities (McCann, 2004).

Strategic agility and resilience capacity share common roots and are built, in part, from complementary capabilities and assets. Moreover, both presume change and surprise can be sources of opportunity. However, they are distinct constructs that are designed to respond to different environmental conditions. Strategic agility is needed to address change that is continuous and relentless while resilience capacity is needed to respond to change that is severely disruptive and surprising (Deevy, 1995; Hamel and Valikangas, 2003; Jamrog et al., 2006; McCann, 2004). Often firms experience both types of change and, t hus, resilience capacity and strategic agility are complementary capabilities that enable organizations to deal with the tumultuous environments in which they operate.

In this chapter we explain how resilience capacity can enable a firm to more fully realize the benefits that disruptive opportunities present and thereby capitalize more fully on its strategic agility. We examine the different forms that strategic agility can take and explain how these different forms lead to different types of outcome. Sustained success depends on a firm's ability to choose the best form of agility for existing strategic purposes and on recognizing the need to change forms as conditions evolve. We argue that resilience capacity contributes to both these decisions.

Resilience capacity is a multidimensional, organizational attribute that enables a firm to effectively absorb, respond to and potentially capitalize on disruptive surprises (Hamel and

Valikangas, 2003; Lengnick-Hall and Beck, 2005; McCann, 2004). It provides a foundation of insight, flexibility, and hardiness that makes it possible for a firm to bounce back and often create new ways to flourish when faced with uncertainty and adversity stemming from a discontinuous jolt within its ecosystem. Resilience capacity is embedded in a set of organizational routines and processes by which a firm conceptually orients itself, acts decisively to move forward, and establishes a setting of diversity and adjustable integration that enables it to overcome the potentially debilitating consequences of a disruptive shock (Lengnick-Hall and Beck, 2005). We define resilience capacity as the organizational ability and confidence to act decisively and effectively in response to conditions that are uncertain, surprising, and sufficiently disruptive that they have the potential to jeopardize long-term survival. Resilience capacity is associated with an ability to solve current problems while preserving flexibility. Resilience capacity offers the *potential* for enhancing the organization's capability set as a direct consequence of the response activities. Modest levels of resilience capacity enable a firm to recover from disruptions and resume normal operations, and high levels of resilience capacity can enable a firm to undergo a robust transformation and thereby thrive in part as a result of the adverse events. Recovery is defined as bouncing back or rebounding from environmental disruptions and resuming established levels of performance. Robust transformation, in contrast, is defined as capitalizing on environmental disruptions in ways that create new options and capabilities. Thus organizational resilience represents a continuum of response ranging from survival to recovery to beneficial transformation. The higher the level of resilience capacity the more reasonable it is to expect that an organization will achieve a position toward the robust transformation end of the continuum.

Strategic agility has been defined as "the ability to quickly recognize and seize opportunities, change direction, and avoid collisions" (McCann, 2004: 47), as the ability to "produce the right products at the right place at the right time at the right price" (Roth, 1996: 30), or as "moving quickly, decisively, and effectively in anticipating, initiating and taking advantage of change" (Jamrog et al., 2006: 5). It captures an organization's ability to manage and adjust to continuous change and so is tied to the frequency and tempo of environmental shifts and indicates a firm's nimbleness and quickness. Strategic agility prepares organizations to embrace relentless change by generating a range of resource and capability alternatives; developing skills at aligning, realigning and mobilizing resources; taking resolute action; and removing barriers to change (Brown and Eisenhardt, 1997; D'Aveni, 1994). Since both resilience capacity and strategic agility underscore a firm's need for deliberate and positive activities in the face of changing conditions, there is a strong connection between these two organizational characteristics. However, there are also important distinctions between the two. Table 3.1 highlights some of the important differences between resilience capacity and strategic agility. One goal of this chapter is to offer a more in-depth understanding of these two constructs and of the interactions between them that enable organizations to thrive in dynamic environments.

We begin with an in-depth discussion of resilience capacity and the component capabilities that comprise this construct. This is followed by an examination of strategic agility. The next section explores the ways in which resilience capacity goes beyond enabling an organization to restore its performance after

**Table 3.1    Comparative analysis of resilience capacity and strategic agility**

| Resilience Capacity | Strategic Agility |
|---|---|
| **Environmental Conditions** | **Environmental Conditions** |
| Disruptive change | Continuous change |
| Surprise | |
| **Foundational Components** | **Foundational Components** |
| Cognitive resilience | Flexible resource base |
| Behavioural resilience | Learning aptitude |
| Contextual resilience | Decision-making prowess |
| | Resilience capacity |
| **Strategic Relevance/Usefulness** | **Strategic Relevance/Usefulness** |
| Survival | Responsiveness |
| Restoration | Proactive adjustment |
| Transformation | Change initiation |
| | Strategic supremacy |
| *Valuable capability for responding to unexpected disruptions across varying environmental and operating conditions* | *Valuable capability for achieving fit between action alternatives, resource configurations, market conditions, and strategic intent* |

a crisis, and assists a firm in preparing for continuous change by enhancing organizational agility and helping a firm navigate among the various forms of agility. We conclude with a discussion of research and managerial implications.

## Building Resilience Capacity

Resilience capacity is a multidimensional set of routines, resources, behaviours, capabilities, and mental models that leads to organizational resilience. As indicated previously, organizational resilience is a firm's ability to bounce back and often create new ways to flourish when faced with disruptive conditions. Resilient organizations are able to absorb the impact of environmental disruptions (Meyer, 1982). They are able to withstand anything that comes along and, depending on their resilience capacity, potentially become more hardy and capable as a consequence of effectively responding to disruptive shocks (Lengnick-Hall and Beck, 2005). While resilient organizations are nimble, flexible and agile, not all agile organizations are resilient (Hamel and Valikangas, 2003; Jamrog et al., 2006; McCann, 2004). The primary distinction is the nature of the environmental shifts each organizational capability is designed to address. Organizational resilience capacity prepares organizations to effectively manage disruptive, unexpected and potentially debilitating change by ensuring the means needed for recovery and renewal are available; absorbing shocks and complexity; broadly accessing resources; crafting creative alternatives; and executing transformational change (McCann, 2004).

## Achieving Resilience Capacity

An organization's resilience capacity is created from interactions among specific cognitive, behavioural, and contextual factors (Lengnick-Hall and Beck, 2005). The mental processes and conceptual orientation known as *cognitive resilience* enables an organization to notice, interpret, analyze, and formulate responses to unfamiliar evolving situations. Cognitive resilience contributes to the generation and selection of action alternatives and to a firm's decisiveness in initiating activities. *Behavioural resilience*, the honed and rehearsed actions that become part of a firm's innate reaction to disruptive conditions, drives the development of particular routines, resource configurations, and interaction patterns that implement the firm's response. These behaviours are designed to both create and capitalize on a firm's flexibility. *Contextual resilience* describes the network of interactions and resources that provide the backdrop for a firm's response to disruptive conditions. Contextual resilience combines interpersonal relationships that provide a foundation for rapid responses to emerging conditions and a network of potential resource donors that enlarges the range of viable options and resource combinations that a firm can consider under disruptive conditions. These three dimensions (cognitive resilience, behavioural resilience, and contextual resilience) play distinct but complementary roles in generating organizational responses to disruption.

These three dimensions of resilience capacity work both independently and interactively to recognize and respond to disruptive change. Synergistic and mutually reinforcing interactions among all three dimensions likely offer the greatest potential for constructing unique, competitively superior resource-development capabilities, an appropriately varied range of competencies and the ability to effectively use these assets. The following sections detail the characteristics associated with each resilience capacity dimension. Subsequent discussion then links the three components of resilience capacity to the creation of particular variations in strategic agility and to the selection and implementation of an effective agility portfolio.

**Cognitive Resilience**

The first dimension of resilience capacity – *cognitive resilience* – is an organizational capability that enables a firm to notice shifts, interpret unfamiliar situations, analyze options, and figure out how to respond to conditions that are disruptive, uncertain, surprising and have the potential to jeopardize the organization's long-term survival (Lengnick-Hall and Beck, 2005). Multiple factors contribute to the creation of cognitive resilience but two of the most important elements are a strong identity and constructive sensemaking.

1. *Organizational identity.* Firms can foster a positive, constructive conceptual orientation through a strong sense of purpose, authentic core values, a genuine vision, and a deliberate use of language (Collins and Porras, 1994). For example, the ways in which organizations frame and label environmental issues (e.g., as a problem or an opportunity) influence the types of responses that are generated (Dutton and Jackson, 1987). The labels used to describe an issue affect subsequent behaviours in terms of risk, commitment, engagement, and persistence. Strong core values coupled with a sense of purpose and identity encourage an organization to frame conditions in ways that enable problem-solving and action rather than in ways that lead to either threat rigidity or dysfunctional escalation of commitment.

2. *Sensemaking.* Cognitively resilient firms are adept at sensemaking in order to interpret and provide meaning to unprecedented, situation-specific events and conditions (Thomas et al., 1993; Weick, 1995). Collective sensemaking relies on the language of the organization (i.e., its words, images, and stories) to construct meaning, describe situations, and imply both meaning and emotion. A prevailing vocabulary that implies capability, influence, competence, consistent core values, and a clear sense of direction sets the stage for constructive sensemaking.

Weick (1993) defines wisdom as an attitude taken towards events or conditions that blends caution and confidence in such a way that expertise leads to understanding at the same time that scepticism leads to curiosity and the search for new information. Wisdom relies on knowledge gained through past experience but it does not stop there. Wisdom is the recognition that each situation contains unique features that may be quite subtle but that can be incredibly powerful in shaping consequences, relationships and actions. Therefore, attitudes that promote wisdom contribute to sensemaking and complement other cognitive elements. To achieve wisdom, firms must actively balance contradictory forces. In other words, constructive sensemaking relies on reciprocal information-seeking and meaning ascription.

*Outcomes from Cognitive Resilience*

The mindset that enables a firm to move forward with flexibility is often an intricate blend of expertise, opportunism, creativity, and decisiveness despite uncertainty. If a firm is too bound by conventional answers or precedent, it will have great difficulty conceiving a bold new path. If a firm disregards real constraints it will forge infeasible solutions. Cognitive resilience requires a solid grasp on reality and a relentless desire to question fundamental assumptions that may no longer apply. In addition, alertness or mindfulness that prompts an organization to continuously consider and refine its expectations and perspectives on current functioning enables a firm to more adeptly manage environmental complexities. Cognitive resilience depends on an ability to conceptualize solutions that are both novel and appropriate (Amabile, 1988). In summary, cognitive resilience includes the mental capabilities and the conceptual orientation that provide the intellectual basis for resilience.

## Behavioural Resilience

Behavioural resilience comprises the established behaviours and routines that enable a firm to learn more about a situation, implement new routines, and fully use its resources under conditions that are disruptive, uncertain, surprising, and have the potential to jeopardize the organization's long-term survival. These actions and activities allow organization members to respond collaboratively to environmental threats and challenges in ways that facilitate a stronger and more competent firm (Lengnick-Hall and Beck, 2003). Routines and activities that comprise behavioural resilience are developed through a combination of practiced resourcefulness and counterintuitive action juxtaposed with useful habits and behavioural preparedness. In this way, behavioural resilience results from a dynamic tension between behaviours that foster creativity and unconventional actions, and familiar and well-rehearsed routines that keep an organization grounded and to provide the platform for inventiveness. Combined, these behaviours create centrifugal forces (influences that make ideas, knowledge and information available for creative action) and centripetal forces (influences that direct inputs and processes toward actionable solutions) that enable a firm to learn more about a situation and to fully use its own resources under conditions that are uncertain and surprising (Sheremata, 2000).

## Resourcefulness

Learned resourcefulness is the accumulation of established and practiced behaviours for innovative problem-solving that result in heightened levels of ingenuity, inventiveness, and bricolage (the imaginative use of materials for previously unintended purposes). As organizations develop and reinforce routines that proliferate ideas, manage conflict to cope with several new ideas at the same time, facilitate change, and initiate novel activities (Kirton, 1976), individuals and organizations become adept at engaging in disciplined creativity leading to unconventional, yet robust, responses to unprecedented challenges (Lengnick-Hall and Lengnick-Hall, 2003; Mallak, 1998a; Weick, 1993). Resourceful behaviours typically combine innovation and decisiveness to capitalize on an immediate situation. Organizations that develop and rehearse behavioural routines that promote resourcefulness and creativity are able to take whatever resources and opportunities are at hand to move the firm forward. Coutu (2002) described these behaviours as "ritualized ingenuity." This can lead to timing advantages including the ability to capitalize on rapid response opportunities, to do more with less, and to use all of a firm's assets to full advantage.

As organizations attempt and succeed at bold, innovative moves, they develop both expertise and confidence. The expertise builds the behavioural repertoire and the confidence builds cognitive resilience. Specific skills and competencies that lead to learned resourcefulness improve with experience and practice (Eisenhardt and Tabrizi, 1995; Senge et al., 1994). For example, divergent thinking skills can be honed through brainstorming, devil's advocacy techniques, and dialogue. Similarly, problem-solving techniques that rely on frequent iterations serve as catalysts for new ideas and increase the odds of success simply because there are more options available for consideration. These behaviours can become familiar as they are applied routinely to solve problems.

*Counterintuitive Moves*

In a study of hospitals dealing with the sudden and unprecedented jolt of striking physicians, Meyer (1982) found that resilient hospitals chose a variety of different paths but one commonality was that the resilient choices were counterintuitive given each of the hospital's normal operating habits. For example, Meyer found that the hospital which typically adopted an entrepreneurial prospector strategy responded to the disruption by centralizing authority, reducing staff and containing costs. Therefore, it appears that a second behavioural pattern contributing to resilience is the ability to follow a dramatically different course of action from that which is the norm for the organization. One way to expand the number of available options is to design a range of strategy assortments that evolve over time and capitalize on uncertainty (Beinhocker, 1999). Behaviours that initiate counterintuitive actions and allow firms to change direction can be practiced to develop increased organizational agility. The more frequently an organization engages in actions that challenge the prevailing status quo, the more adept it is likely to become at quickly and effectively developing a counterintuitive and varied action repertoire.

*Useful Habits*

Third, in direct contrast to learned resourcefulness and counterintuitive action, behavioural resilience also depends on useful, practical habits especially repetitive, well-rehearsed routines that provide the first response to any unexpected threat. Useful habits emerge from genuine organizational values. A cohesive sense of what a company believes (the core values that contribute to cognitive resilience) is the foundation for developing day-to-day behaviours that translate intended strategies into actions. If an organization develops values that lead to habits of investigation rather than assumption, routines of collaboration rather than antagonism, and traditions of flexibility rather than rigidity, it is more likely to intuitively behave in ways that open the system and generate resilient responses.

*Preparedness*

Fourth, behavioural preparedness helps bridge the gap between the divergent forces of learned resourcefulness and counterintuitive action and the convergent forces of useful habits. Behavioural preparedness is taking actions and making investments before they are needed to ensure that an organization is able to benefit from situations that emerge. Behavioural preparedness is the activity-based foundation for informed opportunism (Waterman, 1987). Behavioural preparedness also means that an organization deliberately unlearns obsolete information or dysfunctional heuristics (Crossan et al., 1999; Hammonds, 2002). It is just as important for organizations to quickly discard behaviours that result in inappropriate constraints as it is for them to develop new competencies. Behavioural preparedness enables an organization to act in response to opportunities that other firms without their competencies might forego. Firms that have not developed the necessary behaviours before they are needed jeopardize behavioural resilience because they are unable to capitalize on unanticipated changes in technology, ideas, or market conditions.

*Outcomes from Behavioural Resilience*

Behavioural resilience translates the thoughts and perceptions identified through cognitive resilience into tangible actions and responses. This leads to two important outcomes. First, a combination of learned resourcefulness and counterintuitive actions generates a complex and varied inventory of potential strategic actions that can be drawn upon in emerging situations. Resourcefulness and mobile resources combine to create a reservoir of options that expand the range of possible future behaviours (Ferrier et al., 1999). Second, a combination of useful habits and behavioural preparedness creates a foundation of rehearsed and habitual expert routines that ensure an organization's initial and intuitive action response to any situation will create options rather than constraints.

Drawing an example from the military, patrolling is the most basic, standard, and over-learned routine of the infantry (Simons, 1997). Patrolling is crucial for conducting surveillance or for setting up an ambush or a diversion. In addition, patrolling alerts team leaders to potential deficiencies of individual soldiers and enables the team to rehearse coordination until it becomes ordinary and consistent. These convergent routines can operate as simple rules that create a behavioural gyroscope for guiding organizational actions in uncertain circumstances.

## Contextual Resilience

Contextual resilience provides a setting to nurture attitudes and facilitate behaviours that promote a collaborative response to environmental complexities (Lengnick-Hall and Beck, 2003). Contextual resilience is the combination of interpersonal connections, resource stocks, and supply lines that provides the foundation for quick action under emerging conditions that are disruptive, uncertain, surprising, and have the potential to jeopardize the organization's long-term survival. Factors that contribute to contextual resilience include deep social capital, broad resource networks, and deference to expertise.

*Deep Social Capital*

First, deep social capital evolves from respectful interactions within an organizational community (Ireland et al., 2002). Respectful interactions are defined as face-to-face, ongoing dialogues rooted in trust, honesty and self-respect (Weick, 1993). Respectful interaction builds informed and disclosure-oriented intimacy and is a key factor enabling the collaborative sensemaking component of cognitive resilience.

Deep social capital offers a number of important contextual benefits (Adler and Kwon, 2000). It facilitates growth in intellectual capital since people are more likely and more able to share tacit information. It lends itself to resource exchange since groups come to recognize their interdependence. Social capital also promotes cross-functional collaboration since people appreciate perspectives that are different from their own. Deep social capital is a foundation for exchanges that endure beyond immediate transactions and grow into mutually beneficial, multifaceted, long-term partnerships. Finally, deep social capital can enable an organization to build bridges that cross conventional internal and external boundaries and forge a network of support and resources.

*Broad Resource Networks*

Second, access to broad resource networks is a key element in creating contextual resilience. Resilient individuals are distinguished by their ability to forge relationships with others who could share key resources (Werner and Smith, 2001). Likewise, resilient firms are able to utilize relationships with supplier contacts, loyal customers, and strategic alliance partners to secure needed resources to support adaptive initiatives. Resources gained through a firm's network of organizational relationships generate contextual resilience in several ways. The ability to obtain resources externally tends to ensure some measure of continuous slack. Continuous slack has been found to be more significant than resource abundance in increasing innovation and resourcefulness (Judge et al., 1997) and thus contributes to developing an action inventory. In addition, external resources are likely to extend the range of feasible actions and promote an assortment of alternative applications of these resources. This, in turn, stimulates innovation and challenges prevailing assumptions in ways that can cultivate wisdom. External resources also ensure that bonds with various environmental agents are maintained, thereby reinforcing social capital beyond the firm's boundaries.

*Deference to Expertise*

Deference to expertise is the third factor associated with contextual resilience (Weick and Sutcliffe, 2001). Resilient organizations are not typically hierarchical. Instead, they rely on self-organization, dispersed influence, individual and group accountability, and similar factors that create a "holographic" structure (Morgan, 1997). In holographic structures, each part is a fractional replica of the whole organization. Holographic structures contain systematic redundancy in both information processing and crucial skills to enhance flexibility. They use the minimum specifications possible to ensure collaboration, but leave freedom for experimentation and self-organization. Thus, holographic structures are designed to learn and to change their behaviours based on new insights and information. In addition to relying on these structural designs, resilient organizations share decision-making widely (Mallak, 1998b). Each organization member has both the discretion and the responsibility for ensuring attainment of organizational interests. Overall, this shared responsibility coupled with interdependence creates a setting that facilitates cognitive and behavioural resilience.

*Outcomes from Contextual Resilience*

Contextual resilience establishes the operational platform to facilitate resilient behaviours and attitudes. While contextual resilience is not sufficient to create resilience capacity, it is an integral ingredient enabling the kinds of behaviours and mental models that lead to organizational resilience. Moreover, contextual resilience provides the necessary medium for brewing the other two dimensions of resilience capacity. Without the conduit of relationships, processes, and intangible assets that generate contextual resilience, there would be few ways to synthesize resilient cognitions and behaviours into an enterprise-wide capability.

In summary, cognitive resilience, behavioural resilience, and contextual resilience work together to create an organizational capability that has important implications for the development of strategic agility. Before discussing these implications, however, it is important to examine key aspects of strategic agility.

## Perspectives on Strategic Agility

Strategic agility means that an organization can take quick, decisive, and effective actions and that it can trigger, anticipate, and take advantage of change (Doz and Kosonen, 2007; Jamrog et al., 2006). Firms demonstrating strong agility are able to maintain their strategic supremacy despite market fluctuations (D'Aveni, 1999; Thomas, 1996). A high level of strategic agility means that a firm is able to demonstrate a consistent capacity for concentrating resources on key strategic issues, accumulating new resources efficiently and effectively, complementing and combining resources in new ways, and redeploying resources for new uses (Hamel and Prahalad, 1993). In many ways, strategic agility captures a firm's prowess for developing and learning complex problem-defining and problem-solving heuristics (Lei et al., 1996).

The competitive dynamics literature argues that agility is correlated with a number of factors such as response speed, fast directional changes, number of strategic moves taken in a time period, variety in strategic moves undertaken, a firm's ability to initiate new action sequences, and similar indicators of a broad action repertoire coupled with decisiveness (Ferrier, 2001; Ferrier et al., 1999; Grimm et al., 2006). Much of what we know about how to achieve strategic agility is drawn from our understanding of organizational change (Brown and Eisenhardt, 1997; Goldman et al., 1995; Rindova and Kotha, 2001), exploration and exploitation (Benner and Tushman, 2003; March, 1991; O'Reilly and Tushman, 2004), and dynamic capabilities (Eisenhardt and Martin, 2000; Teece et al., 1997; Winter, 2003).

However, strategic agility can be realized through different component routines, resources, and competencies depending on the conditions and outcomes that a firm is striving to achieve. In other words, the elements of strategic agility that appear crucial in an extremely turbulent and unpredictable market appear to be different than the components of agility that are essential in a more moderately dynamic marketplace characterized by punctuated equilibrium. Moreover, agility that is designed to augment and capitalize on existing sources of competitive advantage is quite different than agility that is designed to result in discontinuous and radically different sources of advantage. A firm with a rich and varied agility repertoire is able to develop resources and competencies that allow effective responses to a range of market and strategic conditions. We discuss the origins of these differences next.

## Strategic Agility and Market Turbulence

The level of market turbulence determines the *pattern* of routines, capabilities, and resource deployments that is likely to be most effective. In environments that are only moderately unsettled, and characterized by punctuated equilibrium, agility is best achieved by patterns that are "complicated, detailed, analytic processes that rely extensively on existing knowledge and linear execution to produce predictable outcomes" (Eisenhardt and Martin, 2000: 1106). Because a firm has a baseline understanding of external conditions, it is able to emphasize *complexity reduction* and focus its analysis on anticipating and understanding the nature, direction, and consequences of the changes that are taking place. Complexity reduction is described as an organizational approach that relies on specialization, abstraction, and codification to devise a single best representation of the environment to which a firm can then adapt in a systematic way (Boisot and Child, 1999). On the other hand, high-velocity, exceptionally turbulent markets require patterns of activity that are much more emergent and fluid. According to Eisenhardt and Martin (2000), agility is best achieved in highly turbulent environments by patterns of behaviour that are "simple, experiential, unstable processes that rely on quickly created new knowledge and iterative execution to produce adaptive, but unpredictable outcomes" (p. 1106). Routines and capabilities for tumultuous markets are designed to *absorb complexity*. Complexity absorption extends the range of environmental contingencies that can be handled by simultaneously considering a variety of sometimes conflicting representations of the environment and by maintaining a broad repertoire of potential actions that could be applied conditionally to meet particular needs (Boisot and Child, 1999).

## Strategic Agility and Prevailing Sources of Advantage

We argue that another crucial difference for the design of activities to achieve strategic agility is whether routines are intended to create new resources and competencies that build on the firm's current configuration, or whether they are intended to create new action patterns that disregard current strengths and work to redefine market value. If a firm determines that superior performance will result from its ability to develop, use, and protect its platform competencies and resources, then its strategy will emphasize *sustaining technologies* and business strategies relying on complementary shifts (Christensen, 1997). Under these conditions, agility will be directed toward a competence-enhancing strategic intent (D'Aveni, 1999; Hamel and Prahalad, 1994). If, on the other hand, a firm determines that superior performance comes from rapidly and repeatedly disrupting the current market situation to create unprecedented and unconventional sources of value, then *disruptive technologies* will underpin its strategic activities (Christensen, 1997; D'Aveni, 1999; Grimm et al., 2006). Under these latter conditions agility will be directed toward a competence-destroying strategic agenda.

## Four Forms of Strategic Agility

As illustrated in Figure 3.1, strategic agility can take a variety of forms which are designed for different market conditions and different strategic purposes. Form 1 (complementary augmentation)

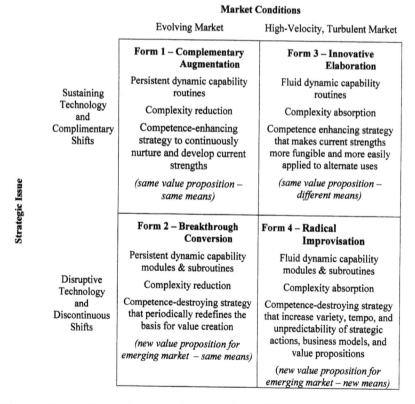

|  | | **Market Conditions** | |
|  | | Evolving Market | High-Velocity, Turbulent Market |
|  |  | **Form 1 – Complementary Augmentation** | **Form 3 – Innovative Elaboration** |
| **Strategic Issue** | Sustaining Technology and Complimentary Shifts | Persistent dynamic capability routines<br><br>Complexity reduction<br><br>Competence-enhancing strategy to continuously nurture and develop current strengths<br><br>*(same value proposition – same means)* | Fluid dynamic capability routines<br><br>Complexity absorption<br><br>Competence enhancing strategy that makes current strengths more fungible and more easily applied to alternate uses<br><br>*(same value proposition – different means)* |
|  | Disruptive Technology and Discontinuous Shifts | **Form 2 – Breakthrough Conversion**<br><br>Persistent dynamic capability modules & subroutines<br><br>Complexity reduction<br><br>Competence-destroying strategy that periodically redefines the basis for value creation<br><br>*(new value proposition for emerging market – same means)* | **Form 4 – Radical Improvisation**<br><br>Fluid dynamic capability modules & subroutines<br><br>Complexity absorption<br><br>Competence-destroying strategy that increase variety, tempo, and unpredictability of strategic actions, business models, and value propositions<br><br>*(new value proposition for emerging market – new means)* |

Figure 3.1    Four forms of strategic agility

and Form 2 (breakthrough conversion) can be achieved through dynamic capabilities and routines that are familiar and rehearsed, capture expertise that has been developed over time, and reflect intricate analysis, planning, and implementation sequences. Form 3 (innovative elaboration) and Form 4 (radical improvisation) can be achieved through dynamic capabilities and routines that are developed in an emergent fashion, are guided by simple rules, and are designed to absorb complexity. Both complementary augmentation (Form 1) and innovative elaboration (Form 3) build on a firm's sustaining technologies and reinforce or apply current strengths. The purpose of these forms of strategic agility is to augment and extend established organizational competencies. In contrast, both breakthrough conversion (Form 2) and radical improvisation (Form 4) emphasize disruptive technologies and reciprocally trigger and respond quickly to discontinuous shifts in the marketplace. The intent of these latter forms of strategic agility is more akin to creative destruction in which existing competencies are unlearned and replaced by new and very different capabilities. An important issue for a firm is choosing the best form of strategic agility for existing strategic needs and recognizing the need to change forms as conditions shift.

An organization's need for strategic agility is directly tied to the rate and persistence of change that the firm encounters. As change becomes increasingly relentless, agility becomes essential for organizational success. Several factors underpin the overarching agility capability regardless of which form is being enacted: (a) a unified managerial commitment; (b) strategic acuity enabling key leaders to identify and appreciate opportunities and threats; (c) fluid and tinkerable resources that can be mobilized, reassembled, and redeployed to meet differing needs; and (d) adept learning, unlearning and knowledge exploitation capabilities (Doz and Kosonen, 2007; Ghemawat and del Sol, 1998; McCann, 2004; Roth, 1996). Different dynamic capabilities, a choice between complexity reduction and complexity absorption, and an emphasis on competence-enhancing versus competence-destroying investments are then overlaid on these foundation factors to create different forms of agility to respond to market conditions and the kind of shifts that must be managed. Over time, an organization may develop a portfolio of different agility approaches to correspond to the different competitive realities it experiences.

## Resilience Capacity and Strategic Agility

Resilience capacity offers firms the potential both to develop strategic agility in a way that matches prevailing environmental conditions and competitive realities, and also create a platform for developing variations in forms of strategic agility over time. This is similar to the role resilience capacity plays in enabling firms to choose between adaptive fit and robust transformation when faced with strong environmental shifts (Lengnick-Hall and Beck, 2005). The logic is fairly straightforward. Resilience capacity stimulates a firm to develop a diverse repertoire of routines and resources. This variety means that a firm is able to construct an array of different combinations of activities and assets to achieve strategic agility. Behavioural and contextual resilience provide the elements, relationships, patterns, experiences, and subroutines that can be mixed and matched to establish competence-enhancing or competence-destroying activities.

Similar to the way that absorptive capacity underpins a firm's ability to appreciate, transform, and exploit new knowledge for strategic purposes (Zahra and George, 2002), resilience capacity underlies a firm's ability to take actions to effectively reconfigure and augment a firm's resources and routines. In addition, resilience capacity captures an important conceptual diagnostic and interpretation component that enables a firm to accurately determine the most appropriate form of strategic agility to use in the current situation. The following sections explain how each of the components of resilience capacity can contribute to various forms of strategic agility.

## Cognitive Resilience and Strategic Agility

The organization-specific routines a firm develops are grounded in the collective consciousness of organization members (Fiol and Lyles, 1985). Therefore, cognitive resilience facilitates a firm's ability to both envision different types of routines that might be needed and to craft specific kinds of routines to respond to particular conditions. In addition, cognitive resilience promotes higher-level learning (Tripsas and Gavetti, 2000) which, in turn, encourages the development of multiple routines and prompts a firm to question its prevailing assumptions. Having a range of different routines and perspectives available to choose from enables a firm to respond to a variety of problem-solving requirements (Lei et al., 1996). Cognitive resilience provides several specific contributions to achieving strategic agility.

First, since cognitive resilience helps members of an organization to see new patterns and consider alternate conditions, it heightens an organization's ability to perceive shifts in the external environment. Moreover, cognitive resilience helps organization members to determine whether environmental changes are temporary or long-standing and whether they are evolutionary or discontinuous (Lengnick-Hall and Beck, 2005). In this way, cognitive resilience leads to an understanding of the environment that helps a firm decide whether a competence-enhancing route or competence-destroying initiatives should be pursued. It also helps a firm to determine whether the most effective strategic actions will be repetitive or emergent. Accurate assessment of the environment is a precondition for selecting an appropriate form of strategic agility.

Second, cognitive resilience contributes to the realization of various forms of strategic agility. Cognitive resilience is useful for refining persistent, repetitive dynamic capabilities and for creating fluid, emerging agility routines. On the one hand, an emphasis on realistic appraisal that underpins cognitive resilience enhances a firm's ability to set challenging but achievable goals and to structure internal processes for maximum feasibility, efficiency, and scalability. Effective goal setting and efficient process design provide a foundation for learning and refining routines over time and developing recurring patterns of behaviour. On the other hand, insight and wisdom emerging from applied cognitive resilience is a foundation for creating and selecting the simple rules that drive more fluid forms of strategic agility. Cognitive resilience also constrains pressure to undertake either gratuitous invention or reckless initiatives. Thus, cognitive resilience offers a useful constraint on investment under uncertainty.

Third, the conceptual skills that enable cognitive resilience are also crucial for envisioning effective alternatives regardless of which form of strategic agility is selected. In this way, conceptual skills that lead to resilience capacity concurrently contribute to agile resources and routines. Gillette offers an example of cognitive resilience in action. When Bic introduced the disposable razor it drastically redefined the shaving industry, cannibalizing Gillette's cartridge system, and switching the primary signal of value toward price competition (D'Aveni, 1999). Gillette faced three choices: (1) elaborating its current business model (competence-enhancing); (2) following new rules that required entirely different resources and competencies from those the firm possessed (competence-destroying by following a rival); or (3) using its resources and competencies in new ways to redefine the industry again (competence-destroying and setting the agenda). Gillette realized it could not compete effectively using Bic's ground rules, and that its current business model was on shaky ground, so they chose to create a further discontinuous shift in the market by redefining value to shaving quality and brand image. They also developed the capacity to periodically punctuate the market with disruptive shifts in order to maintain their strategic supremacy (D'Aveni, 1999). Cognitive resilience enabled accurate diagnosis and effective choice.

Finally, cognitive resilience plays a role in selecting between complexity reduction and complexity absorption. As indicated previously, cognitive resilience increases the probability that a firm will be able to accurately distinguish between temporary, permanent, and continuous changes in their external environment (Lengnick-Hall and Beck, 2005). Thus, cognitive resilience enables a firm to determine whether analysis can lead to the discovery of a single preferred solution or whether the firm must maintain an understanding of competing influences and multiple potential interpretations. If it is possible to design a single best answer, then complexity reduction can be employed. However, if conditions are perpetually emerging, then a complexity absorption strategy is more effective. Cognitive resilience is particularly useful with complexity absorption approaches because the same types of skills that enable individuals and organizations to balance competing forces and achieve wisdom and sensemaking also allow people and units to hold multiple perspectives simultaneously.

## Behavioural Resilience and Strategic Agility

The components of behavioural resilience have a strong influence on both persistent and fluid capability routines. First, useful habits are vital for rehearsing and honing the analytic processes that efficiently use existing knowledge. Useful habits also facilitate linear execution of established work processes to produce the predictable outcomes associated with persistent dynamic capabilities. Second, a deliberate strategy of frequent, time-triggered, diverse competitive moves is a key element in developing a complex action repertoire (Smith, Ferrier, and Grimm, 2001). This type of premeditated, tightly-orchestrated routine is a prototype for complementary augmentation and for breakthrough conversion forms of agility. Third, a complex and varied action inventory provides a rich repertoire of alternatives for responding to the emerging, iterative requirements for the fluid dynamic capabilities needed for innovative elaboration and radical improvisation. Interdependence makes it easier for organizations to recognize the value of ideas that come from other sources and to see the benefits of mixing and matching resources in unprecedented ways. Successful emergency response and disaster recovery organizations, for example, often rely on the useful habit of modular work teams, incident command, and project-based assignments to reinforce and rehearse their rapid deployment capabilities. For these firms, the lessons learned during familiar assignments lead to the expertise needed for unprecedented and unconventional missions.

Likewise the components of behavioural resilience can contribute to either a competence-enhancing or competence-destroying strategic orientation. Useful habits reinforce the core competencies that drive a competence-enhancing strategy and provide a focus for accumulating and concentrating resources. A complex and varied action inventory provides the raw materials necessary to execute unconventional experiments, disconnected simultaneous actions, and the quick responses that are associated with competence-destroying strategies. The factors that contribute to behavioural preparedness also stimulate a firm's ability to vacillate between competence-enhancing and competence-destroying actions. Effectiveness at unlearning coupled with a willingness to make prerequisite investments in assets, human capital, or various capabilities insures that the ingredients are available for either type of strategic initiative (Kogut and Zander, 1996).

Complexity reduction requires codification (assigning data to categories) and abstraction (limiting the number of categories to which data can be assigned) (Boisot and Child, 1999). As firms repeatedly follow organization-specific rules for evaluating, sorting, and analyzing data they become more adept at complexity reduction. In addition, as a firm elaborates its action inventory it increases the likelihood that the means to implement a wider variety of inventive approaches will be available. This increases opportunities for complexity absorption by expanding the number of contingent conditions that can be accommodated and improves complexity reduction by increasing the likelihood that any given alternative identified as ideal will be feasible to implement. However, as a firm increases the variety of its action inventory, it increases the number of diverse elements in the system, which raises the level of complexity within the firm. Complexity is often accompanied by increased specialization. This, in turn, can lead to complicated routines designed to promote higher-order learning and reduce complexity (Lei et al., 1996). Alternatively, as complexity increases, firms can become more comfortable with ambiguity and develop routines that encourage both exploitive and exploratory learning (March and Levinthal, 1993), thus enhancing their ability to absorb complexity. In this way a complex and varied action inventory provides the ingredients for generating both complexity reduction and complexity absorption routines.

## Contextual Resilience and Strategic Agility

Deep social capital, broad resource networks, and deference to expertise, the elements comprising contextual resilience, also shape the dimensions leading to different types of strategic agility. The repeated interactions typical of deep social capital provide the connections necessary to fine tune the persistent dynamic capabilities associated with complementary augmentation and with breakthrough conversion. Deep social capital provides a level of trust and commitment to community (rather than individual) interests that encourage learning from experience and from unexpected surprises. This kind of learning is essential for a firm to benefit from the iterations of fluid dynamic capabilities associated with innovative elaboration and radical improvisation.

Deep social capital also provides a support system that enables a firm to engage in unfamiliar and unconventional activities and to learn from failure. An ability to accommodate mistakes is a prerequisite for competence-destroying strategies since some failures are inevitable and even successful strategies are short-lived. Broad resource networks help firms figure out how best to use knowledge by offering an array of alternatives, and enable timely information exchange from increased connectivity. These factors can facilitate either competence-enhancing or competence-destroying capabilities. In addition, interpersonal networks facilitate the implementation of sustaining technologies by reducing turnover, encouraging collective goals, providing clear and transparent reward criteria needed to encourage collaboration, and forming a basis for reconciling differences (Inkpen and Tsang, 2005). Finally, shared responsibility and interdependence make it more likely that diverse perspectives will be heard and considered. This increases the likelihood that an effective strategic orientation will be selected.

Contextual resilience also influences complexity reduction and complexity absorption. Stable personal ties make it easier to develop norms and rules that govern complexity reduction activities (Inkpen and Tsang, 2005) and to construct the organization-specific language needed for efficient codification (Boisot and Child, 1999). Complexity reduction depends on complementary external and internal actions and deep social capital facilitates the search for common ground. The trust and social support resulting from deep social capital helps buffer the stress of complexity absorption. Moreover, resources secured through opportunistic exchanges with outside groups are a crucial factor enabling complexity absorption (Boisot and Child, 1999).

## Resilience Capacity, Strategic Agility and Organization

*Performance Under Challenging Conditions*

Three important relationships between resilience capacity and strategic agility have been discussed in the previous sections. These proposed relationships are depicted in Figure 3.2.

As indicated previously, resilience capacity and strategic agility reflect a number of common roots including the need for change and emergent behaviour; creativity; intentional, purposeful decision-making and action; and requirements to act despite uncertainty. Consequently, many of the building-block skills, resources and competencies that contribute to resilience capacity simultaneously help develop a firm's strategic agility. For example, creative problem-solving routines, a degree of organizational slack, a clear sense of purpose, high levels of intellectual and social capital, and a propensity for iterative, double-loop learning contribute to both the development of strategic agility and the development of resilience capacity. Therefore, whether a firm considers its current need for resilience

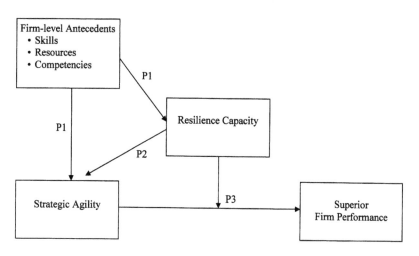

**Figure 3.2    Interactions among resilience capacity, strategic agility, and organizational performance**

capacity to be extremely high or relatively modest, investments made to develop this capacity are quite fungible. Once the skills, resources, and competencies are in place they can be applied to both resilience capacity and strategic agility allowing a firm to leverage its investments in a highly productive manner and to prepare for whatever the firm encounters. Proposition 1 asserts that common investments can be used to build resilience capacity and strategic agility.

*Proposition #1*: Many of the skills and competencies that contribute to resilience capacity also contribute to strategic agility; therefore, as a firm works to develop its resilience capacity it concurrently creates a foundation for agility.

There are numerous interactions among the three components of resilience capacity and a firm's ability to develop and select an effective strategic agility portfolio. For example, the empowering interpretation of the world and self-efficacy that accompanies cognitive resilience enables a firm to act on its decisions despite uncertainty and complexity. Similarly, a complex and varied action inventory increases a firm's absorptive capacity because it has developed expertise in a broader range of activities. This, in turn, increases the firm's ability to recognize value in new knowledge, which leads to enhanced cognitive resilience. The perspective and mental agility that stem from cognitive resilience provide a foundation for a firm to be able to learn from the consequences of the actions it undertakes within its complex action repertoire. Useful habits such as continuous dialog and the trust that results from deep social capital provide the raw material for constructing meaning and making difficult choices in ambiguous situations. Proposition 2 explains how the direct and indirect connections between the two constructs suggest that resilience capacity is not only the basis for restoring a firm's performance following a crisis, but can be a foundation for developing strategic agility as well:

*Proposition #2*: Firms with high levels of resilience capacity are more likely to have a more robust and diversified agility repertoire than is available to firms with little resilience capacity.

Strong resilience capacity creates a useful internal guidance system for organizational analysis and decision-making. The outcomes of cognitive resilience enable a firm to more accurately diagnose environmental conditions and to select the most effective strategic posture in terms of building upon current sources of advantage or creating fundamentally different sources of advantage. As Eisenhardt and Martin (2000) point out, effective application of dynamic activities requires both ingredients and a recipe. The varied action repertoire that is the result of the divergent forces of behavioural resilience provides the ingredients needed to apply strategic agility for competitive advantage. Simultaneously, the convergent forces contributing to behavioural resilience (useful habits and behavioural preparedness) often yield simple rules to guide organization choices under turbulent conditions. Simple rules provide an effective recipe for leveraging the new resource and capability ingredients that an organization produces. Finally contextual resilience offers fertile ground for using strategic agility to best advantage. Together these implications lead to three propositions regarding the ways in which resilience capacity moderates the relationship between strategic agility and organizational performance and helps translate preparation into realized success.

*Proposition #3a:* Resilience capacity strengthens the relationship between a firm's strategic agility and its performance by facilitating the selection of an appropriate form of agility for use given the existing environmental conditions and strategic orientation.

*Proposition #3b:* Resilience capacity strengthens the relationship between a firm's strategic agility and its performance by engaging those behaviours that are particularly useful for applying agility to a specific strategic orientation.

*Proposition #3c:* Resilience capacity strengthens the relationship between a firm's strategic agility and its performance by harnessing and capitalizing on networks and complementary external resources.

## Discussion and Conclusions

Three themes underpin the ideas presented in this chapter:

1. Resilience capacity and strategic agility rely on complementary resources, skills, and competencies. Consequently, as a firm

builds its resilience capacity it simultaneously develops a foundation for creating strategic agility.

2. Strategic agility is a complex, varied construct that can take multiple forms. The likelihood of achieving desirable competitive results increases if the particular form of agility is a solid match to the degree of market turbulence and to the nature of the shifts taking place.

3. Resilience capacity can substantially contribute to a firm's ability to both develop a robust strategic agility portfolio and aid in the selection of the most appropriate form of strategic agility for a particular strategic condition. Therefore, resilience capacity can be viewed as a moderator of the relationship between a firm's strategic agility and subsequent performance.

One important contribution of this chapter is a better understanding of the relationship between resilience capacity and a firm's ability to develop the different forms of agility that enable it to thrive over time and under diverse conditions. A better understanding of this relationship suggests a number of interesting research directions. For example, while a variety of resources and competencies are likely to underpin both strategic agility and resilience capacity, it would be useful to examine which specific resources are universally useful in generating both attributes and which resources and competencies are more strongly associated with resilience capacity or specific forms of strategic agility. Similarly, it would be beneficial to identify specific resources and competencies that are essential for developing resilience capability and strategic agility and to distinguish crucial assets from those that are beneficial but discretionary. In addition, it would be useful to explore organizational processes that increase or decrease the versatility of resource and competency applications toward both objectives.

A second important contribution of this chapter is a detailed, specific, and potentially measurable description of the components of resilience capacity. Definitions and descriptions that have appeared in previous research are often more vague, and have not discussed the component elements in depth. In this chapter, we articulated the crucial elements that underlie the path-dependent process of creating resilience capacity and described these factors in terms that can be more directly operationalized.

Third, this chapter explains the impact resilience capacity may have in strengthening the performance benefits of a robust strategic agility portfolio. Most research treats strategic agility as a uniform construct. We argue that different types of agility are needed for different competitive conditions. In the same way that the agility necessary for superior performance on a basketball court is fundamentally different from the agility needed to provide emergency care following a tornado, organizations need different forms of agility to take advantage of the different environmental situations and requirements they encounter. For example, organizations responding to relentless product improvements from aggressive competitors within their strategic group require a different form of agility than organizations competing with emerging rivals who are actively redefining their value proposition and introducing unfamiliar technologies. The routines, assumptions, and processes that enable a firm to effectively reconfigure its value proposition are quite different from the resources, beliefs, and approaches that enable it to be flexible in terms of the means it uses to achieve selected objectives. Resilience capacity enhances a firm's ability to select the most appropriate form of agility at a particular point in time. It also provides support that facilitates an organization's efforts to implement, reconfigure, integrate, or release resources toward a desired configuration. Additionally, resilience capacity provides increased access to important resources by fostering and building on strong network relationships. An understanding of the connections among resilience capacity, strategic agility, and competitive performance contributes to the growing literature on intangible assets.

There are also several useful managerial implications arising from this chapter. Most organizations operate under conditions of resource limitations or scarcity. Consequently, investing in resources and competencies that can be effectively leveraged because they can be combined easily with other complementary assets or because they can be applied flexibly for multiple purposes is positively correlated with organizational performance (Hamel and Prahalad, 1993). Recognition that certain types of resources and capabilities contribute to both resilience capacity and strategic agility can help firms develop improved investment strategies. For example, investments in human capital to develop employees who are adept learners, strong communicators, and skilled at creating strong interpersonal ties create a foundation for both resilience capacity and strategic agility. Similarly, developing organizational skills such as "ritualized ingenuity" (Coutu, 2002), temporal pacing (Eisenhardt and Martin, 2000), using action to shape cognition (Weick, 1995), and counterintuitive thinking (Meyer, 1982) contribute to both organizational attributes. Even choices regarding physical resource allocations such as designing buildings with open architecture to facilitate interaction and information systems such as knowledge repositories to increase the stock of ideas available can enable a firm to develop assets that are more effectively leveraged.

Resilience capacity can be developed and managed. This implies that managers should build the capacity to effectively attend to, analyze, and understand environmental conditions by establishing a strong organizational purpose, and communicating this purpose throughout the firm to encourage decision-making and action that is consistent with the firm's core values. In addition, managers should ensure their firms develop the capacity to take successful joint action when they incorporate behavioural routines of resourcefulness and creativity while also identifying and maintaining useful habits in an effort to provide strategic agility. Third, managers should foster their firm's capacity to utilize environmental analysis and implement behavioural routines by establishing settings that are conducive to inter- and intra-organizational relationships. Together, these organizational relationships open access to skills, resources, and competencies useful for improved analysis and greater diversity in behavioural responses to uncertain and surprising conditions. Finally, managers should actively attend to their firm's resilience capacity levels in order to achieve greater strategic potential from their strategic agility.

In conclusion, change is an inevitable feature of organizational life. Sometimes change is mandated by powerful external agents. Sometimes change is the natural consequence of interdependence and interaction. Sometimes change is a deliberate strategic initiative designed to increase competitive advantage. Regardless of the causal trigger, organizations must be able to efficiently and effectively alter their resources, competencies, and business models in order to survive and thrive. When change is imposed on a firm, both resilience capacity and strategic agility are essential for selecting an effective responsive posture and for implementing the transformation. When change is an internal choice, strategic agility may initially take precedence, but it is likely that resilience capacity will play an important role in enabling the firm to make subsequent adjustments in response to the reactions of other firms in its marketplace. Resilience capacity is the basis for building sufficient diversity into a firm's strategic agility to enable a portfolio of options and outcomes. Strategic agility that is deep, broad, and varied has been argued to be the foundation of strategic supremacy (D'Aveni, 1999; Eisenhardt and Martin, 2000; Ferrier, 2001). In summary, if an organization wants to be able to recover from adversity, thrive amid turbulence and environmental jolts, and set an agenda that capitalizes on the inevitability of change, it must develop both resilience capacity to ensure restoration and rejuvenation, and strategic agility to prepare for the adjustments it will need to make when faced with the unpleasant surprises and/or unprecedented opportunities that sometimes accompany relentless change. A better understanding of resilience capacity, strategic agility and the interactions between these two attributes will offer new ways to explain why some firms continue to outperform others.

# PART II
# Models and Measures

# Chapter 4

# An Initial Comparison of Selected Models of System Resilience

David D. Woods
Jason Schenk
Theodore T. Allen

## Introduction

The label "resilience" has come to be used by many researchers in biology, psychology, organizational dynamics, safety, and complexity theory to describe one or another aspect of the adaptive capacity of organisms, species, groups, and organizations. As a result, there are many potential definitions of and representations for resilience, each of which suggests different ways to parameterize a system's adaptive capacity. This chapter compares several proposed models for resilience of human systems in order to determine how they are similar and where there are substantive differences.

The use of resilience to refer to human systems and organizations began with Holling (1973) with respect to ecological systems, and led to the subsequent development of the Social-Ecological Systems (SES) approach (e.g., Gunderson, 2000; Holling et al., 2002; Walker et al., 2004). Others have transferred engineering models such as ball and cup dynamics (Scheffer et al., 1993) or models of physical springs to describe how a system is resilient. This chapter briefly provides an overview of each model in order to make preliminary comparisons. The result is then connected to Rasmussen's proposed Safe Operating Envelope concept used in systems safety (Rasmussen, 1997b), and contrasted with the Stress-Strain State Space proposed by Woods and Wreathall

(2008) which is derived from an analogy to parameter spaces in materials science.

The models compared are models in the sense they represent a representational system that captures key concepts and relationships about how systems can be more or less resilient. The representations tend to use state space diagrams, a common form of representation in engineering, to propose or capture general empirical relationships as parameters in the state space. The models were selected for comparison because the parameters associated with each representation facilitated comparisons.

Reviewing the models allowed us to generate a broad set of concepts about how systems are resilient. The models are then compared by building a matrix of concepts, by models (see Table 4.1). While the model review only focuses on four models that have suggested specific parameters for system resilience, the comparisons can be extended by adding additional concepts and considering additional models that can be considered relevant to resilience (e.g., Hollnagel's functional resonance accident model [2004]).

The comparison reveals two insights. First, the parameters of Holling's ecological systems and ball and cup models are essentially the same and only characterize a system's base adaptive capacity. Second, the Stress-Strain State Space is a broader characterization of a system's different adaptive capacities and the transitions between them. The Stress-Strain State Space incorporates the parameters from the other models, while also providing other parameters that capture many other facets of adaptive capacity, brittleness, and graceful degradation in particular. Finally, we discuss the connection between the Stress-Strain State Space and Carlson and Doyle's (1999; 2002) theoretical framework of highly optimized tolerance in complex systems.

## Ecological Systems

Holling (1973) introduced resilience into the ecological systems literature based on the observation that systems are rarely static at equilibrium. Rather, the fluctuation of the system is the important attribute. Factors such as fecundity and mortality may lead to multiple equilibria, and the transitions between states are of interest. Holling (1973) defines two parameters related to random factors that might affect a system. The first is stability, or what he later terms "engineering resilience," which he defines as the ability of a system to return to equilibrium quickly and with minimal fluctuation after a temporary disturbance. Using a generic linear system, the parameter can be expressed as:

$$\frac{dx}{dt} = Ax \quad (1)$$

which represents a linear, or linearized, system near its equilibrium. Pimm and Lawton (1977) defined $-1/Re(\lambda_1(A))$ as this measure of return to equilibrium, or engineering resilience, where $\lambda_1$ is the eigenvalue (characteristic root) with the largest real part. This value is essentially the slope of slowest descent back into the equilibrium.

Holling defined the second parameter, ecological resilience (originally termed, simply, "resilience") as the ability of a system to absorb change and disturbance and still maintain the relationships between its state variables. For state space models, this point may be specified by transition rules. Table 4.1 summarizes the parameters in Holling's model.

**Social-Ecological Systems**

Holling's work provided a foundation that evolved into the study of Social-Ecological Systems (SES). Walker et al. (2004) defines three attributes that guide the trajectories of an SES. These are

**Table 4.1**  **Ecological systems theory labels related to resilience**

| Original term | Later term | Definition | State Space correlate |
|---|---|---|---|
| Stability | Engineering resilience | Return time to equilibrium | $\dfrac{-1}{Re(\lambda_1(A))}$ |
| Resilience | Ecological resilience | Amount of stress before restructuring | Transition rule |

resilience, adaptability, and transformability. Resilience refers to the systems ability to compensate for disturbances while preserving function. Adaptability refers to the ability to manage the resilience of a system (Folke et al., 2005). Transformability is the related concept of how easily a system can be restructured into a new form or functions.

Resilience is further broken down into four components:

1. Latitude, which is the maximum amount a system can be pushed before changing states.
2. Resistance, or how much the system resists pressures to change.
3. Precariousness, which is a measure of how close the current state of the system is to a state change.
4. Panarchy (Holling et al., 2002), which represents cross-scale interactions (Folke, 2006).

More global systems may operate over wider spans and may manage more global or abstract parameters. More local systems respond more quickly, modifying past behaviours in favour of more short-term opportunities to achieve goals. When balanced, interactions over these levels can produce systems that avoid maladaptive patterns (e.g., variations of the tragedy of the commons). Ostrom (1999) calls such multi-level control "polycentric" (cf. also Woods and Hollnagel, 2006).

## Ball and Cup Dynamics

Scheffer et al. (1993) describe system states using a ball and cup metaphor (see Figure 4.1). Each system state is represented by a basin or a zone of attraction. The current operating point is represented by the ball which may reach equilibrium at the bottom of the basin. The speed at which equilibrium is reached is a factor of the slope of the basin. More specifically, the smallest slope of the basin represents the longest possible time it will take for the system to return to equilibrium. This slope is the equivalent of the engineering resilience parameter that Holling defined. Another way to consider this value is the amount of

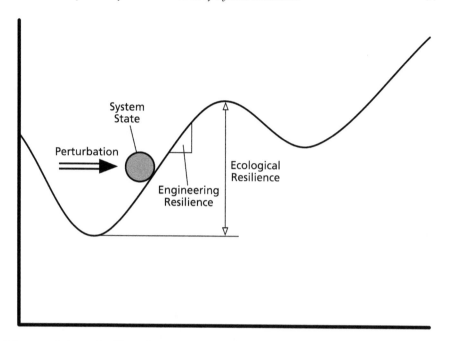

**Figure 4.1   Ball and cup model in two dimensions**

resistance to pressure that the system exerts, which is the concept of "resistance" defined by Walker et al. (2004).

The amount of disturbance that the system can withstand before restructuring itself – the ecological resilience or latitude – is represented by the height of the lowest lip of a basin. Most, if not all, complex systems have multiple zones of attraction (Figure 4.1). This means that if the pressure exceeds the height (or ecological resilience) in a given direction, the system will enter a new state.

## Rasmussen's Safe Operating Envelope

Now let us rotate the three-dimensional ball and cup model to view it from above. We see a basin (or multiple basins) bounded by lips on all sides. The basin represents a "zone of attraction" within which the system maintains a single operational state. The lips represent the boundaries of operation after which the system will move towards a new state. If no new state is available, the system will fall over the edge and fail. Boundaries may be related to economic efficiency, workload, or safety. Figure 4.2 illustrates the classic Safe Operating Envelope diagram (adapted from Cook and Rasmussen, 2005).

The boundaries represent failure points. These are instances in which a system breaks down in terms of economic factors (failures of efficiency or productivity), workload factors (when the workload on people in various roles exceeds time and resources available), or safety factors (when breakdowns occur that injure workers, patients, or the public). Becoming aware that the current system operating point is too close to a boundary, or that the risks of operating that close are too high, leads to adaptive responses. These responses move the system operating point away from that boundary (which could lead the system operating point to move close to another boundary).

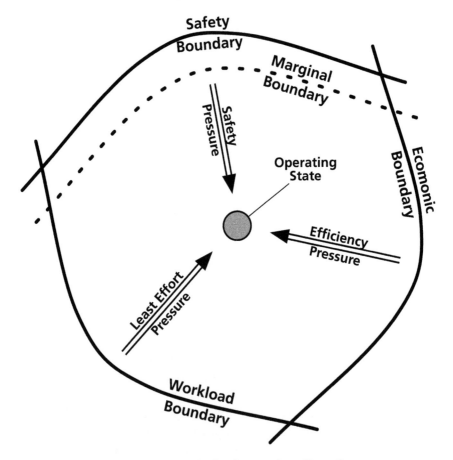

**Figure 4.2      Rasmussen's Safe Operating Envelope**

*Source: Adapted from Cook and Rasmussen, 2005.*

For example, before the Columbia accident, NASA space shuttle operations were under productivity pressure to maintain or improve shuttle turn-around between missions. This was in order to support the space station schedule (i.e., the focus was on moving the system operating point away from the economic failure boundary) (CAIB, 2003). Only after the Columbia accident did it become clear that shuttle operations had been operating too close to the safety boundary with respect to the dangers posed by foam strikes during launch. This also had large effects on mission schedules, which embody NASA's productivity.

As emergency room (ER) personnel experience increasing patient loads and more difficult cases, they are aware that patient care could degrade due to workload bottlenecks. In these circumstances, there may not be enough resources – personnel, expertise, equipment, space – to provide effective, safe, and timely care (Wears, Perry, Anders, and Woods, 2008). Recognizing that an emergency room is approaching workload saturation could lead to a decision to divert incoming patients to another hospital's ER. This risks delays and inadequate care to the patients who are diverted (a safety boundary failure for a subset of patients). Or, in reverse, hospital management might fear consequences for the hospital system should a diverted patient suffer an avoidable bad outcome as a result of the delay. As a result management could adopt a policy that the system's ER will accept all incoming patients. If they did, hospital management would also need to provide the means (criteria, authority, and resources) to expand the workload capabilities of an ER under risk of a workload bottleneck failure. Without the means, accepting more patients increases the risks of a safety breakdown as the ER attempts to cope with the surge (Wears and Woods, 2007).

The Safe Operating Envelope model is based on the concept that goal conflicts, and how organizations respond to the tradeoffs and dilemmas created, are central drivers to the brittleness or resilience of complex systems (Woods et al., 2009; Rasmussen, 1997b; Cook and Rasmussen, 2005; Woods, 2006; Woods and Hollnagel, 2006). In the Safe Operating Envelope model, adaptive behaviour is a function of feedback about growing risks of one or another type of failure as systems undergo change due to new capabilities and intensifying pressures to achieve multiple goals. Parts of the organization that adapt to shift the system's operating point away from one boundary can move the system inadvertently closer to failure with respect to other boundaries. An organization can have a poor model of where it is operating in the space. Where it thinks it is operating can be quite different from where it is actually operating. This gap is what is meant by miscalibration. The system's actual current operating point can be quite different from where the operating point was when the organization was last well-calibrated.

## Stress-Strain State Space

The Stress-Strain or Demand-Stretch State Space model of resilience proposed by Woods and Wreathall (2008) provides the basis for comparison and consolidation with the current alternatives. In materials science, materials can be described in terms of how the material behaves in response to outside demands that place stress on the material. Regularities in the relationship of demands to material behaviour are captured in the different regions and parameters of a stress-strain state space. Woods and Wreathall used this as a point of analogy to consider how systems behave (stretch) in the face of demands imposed by events and variations in the environment.

The Stress-Strain State Space is a kind of fitness space to show how systems adapt to changing demands (see Figure 4.3). As demands increase over some range, systems, such as materials, can stretch uniformly (proportional to the demand). In this first region there is a linear relationship between demand and how the system or organization stretches to accommodate those demands. Performance in this region is characterized by the slope (Young's Modulus) and length, or in combination, Yield Height. These parameters capture the base adaptive capacity of the system in question based on the situations and changes the system can handle in terms of its plans, procedures, training, and personnel capabilities – its on-plan performance area or competence envelope (Woods, 2006).

Should loads on the system continue to grow (the transition point between elastic and plastic regions in materials science) they may exceed or differ from those situations built into the

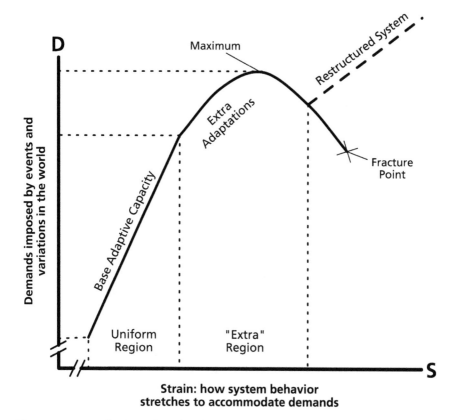

**Figure 4.3**    **Stress-Strain State Space diagram**

*Source: From Woods and Wreathall, 2008.*

competence envelope. This defines the transition to a second region where gaps start to appear and accumulate, moving the system toward failure (a fracture point).

In systems, though, personnel responsible for effective and safe system operation can notice that gaps are occurring. They can start to compensate by actively adapting behaviour strategies and recruiting resources to cope with the new or rising demands. When these extra or second-order forms of adaptive capacity become exhausted, the system can begin to devolve toward a failure point (risk of business collapse, market unravelling, or injuries to workers or the public) or the system can be re-organized to function in a new way. The hospital emergency department is a classic exemplar of a system that is challenged to make transitions across regions in the state space. It transitions from uniform to extra as surges in demand occur, and from extra to restructured, as in mass casualty events (Wears, Perry, Anders, and Woods, 2008).

## Active Reserves and Graceful Degradation

A key challenge to resilience is to maintain and use reserves as sources for adaptation in the extra region without their being dismissed as mere inefficiencies (Woods, 2006). Reserves could be simply redundant resources. For example, staffing for peak demand or designing for maximum expected structural load plus a safety margin. These "brute force" forms of reserve tend to dissipate as "faster, better, cheaper" pressure builds in an organization (Woods, 2005, 2006). Another type of reserve is not maintained online at all times. It is instead used for other purposes and then re-directed; brought to bear when challenging or peak situations arise. An example of this type of active reserve is the use of on-call staff or special response teams. Crisis management preparation and training often focuses on how to bring such teams together and deploy them in an evolving and deteriorating situation (Winters et al., 2007). Holling's concept of the spatial mosaic refers to the interaction and redundancy of local populations to achieve a greater global resilience. If a local population is wiped out, for example, more successful surrounding populations can invade that territory and replace it, making the system more persistent, or resilient. Functional reserves refer to shifts in cognitive and collaborative strategies that occur as demands cascade. Space shuttle mission control demonstrates this type of reserve (Watts-Perotti and Woods, 2007; Watts-Perotti and Woods, forthcoming).

In the Stress-Strain State Space, reserves allow for systems to make shifts across sub-regions as the organization taps different resources, providing additional capacity to stretch in the face of new demands (Woods and Wreathall, 2008). Therefore, a system that uses reserves effectively will demonstrate several successive extra regions. It will also be able to continue to function (stretch) longer even as new or changing forms of demands build up. This is resilience in the sense of having the capacity for second-order (extra) adaptive capacity. The ability to transition from uniform to extra regions depends on setting up and deploying reserves.

Emergency departments (ED) provide an example of how systems adapt in the face of changing demands (Wears and Woods, 2007; Wears, Perry, Anders, and Woods, 2008). As demand (type, number, and difficulty of patient needs) exceeds the capacity of the current staff and physical space, adaptations are made. These can include using hallway beds and foregoing or delaying paperwork and documentation. Increasing demands lead to prioritization decisions as queues build up (capacity shortfalls) and workload bottlenecks occur (delaying documentation tasks). Before increasing load risks violations of important standards of care, a decision may be made to implement a high-volume plan. Such a plan restructures the way the department is run and recruits staff and resources from other departments and on-call staff (e.g., in a mass casualty event). Success lies in the ability to deploy reserves in a timely manner. As Woods and Wreathall noted, resilient systems are those that can make effective state transitions across regions in the Stress-Strain State Space.

Decompensation is a basic pattern in adaptive systems. As demands increase, the system can no longer continue to stretch to accommodate the increasing demands (Woods and Cook, 2006). In decompensation, there is a failure to anticipate or recognize that the system is exhausting its capacity to respond to new or changing demands. There is also a failure to adapt behaviour in ways that could allow the system to continue to function to achieve goals (Woods, forthcoming).

First, a system can have greater or lesser base adaptive capacity (as assessed by several models including Holling's ecological system, and ball and cup models). But all systems have limits on these parameters (both for practical and theoretical constraints). Once demands exceed base adaptive capacity and a system has little to draw on, performance quickly falls toward a failure point when the base capacity is exhausted. When systems have additional reserves, they degrade more gracefully. They can draw on these sources of extra adaptive capacity to continue to handle demands and forestall any collapse toward a failure point, such as in the ED example.

These variations in system performance can be captured directly in the Stress-Strain State Space. Plotting where a system lies in terms of this space captures that system's potential for adaptive action in the future. But estimating the parameters of this state space is carried out by assessing how the system has adapted to classes of past challenge events, assessing what provided sources of extra adaptive capacity, and assessing whether these sources are eroding or growing. In Figure 4.4, the basic Stress-Strain State Space is annotated with several examples of additional parameters that capture different aspects of a system's ability to respond to demands including the graceful degradation margin.

## Model Comparison Criteria

Each of the proposed models of resilience represents a different perspective of the problem of understanding how systems adapt

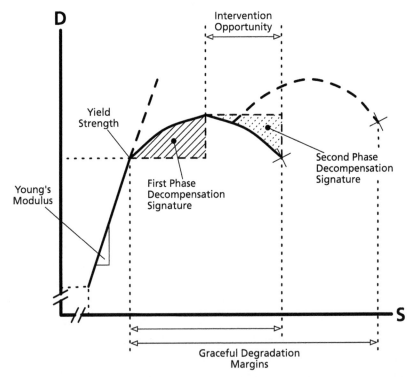

**Figure 4.4**    **Sample of parameters in the Stress-Strain State Space diagram related to issues in system resilience**

to events and change in their environment. To provide a basis for comparison, we extracted a set of general issues, concepts or themes used or addressed in one or more of the proposed models. Note, though, that the comparison itself focuses only on the four models that suggested parameters to assess a systems adaptive capacity. Each row in the set asks a question of the form – "Does the model include concepts about…?" or "How does the model capture or include this concept…?" The set includes:

- what kinds of modes occur in the fitness space proposed?
- parameter(s) that capture system stability in the face of disruption
- parameter(s) that capture the amount of stress that can be handled before restructuring occurs.
- is there a concept precariousness (being near or far from the "edge")?
- how are reserves defined or modelled (e.g., passive or active reserves; deployment preparation)?
- does the model capture the decompensation pattern?
- how does the model define different kinds of adaptive capacity, in particular base or first-order adaptive capacity and second-order adaptive capacity?
- parameter(s) that capture how gracefully system performance degrades under stress.
- how does the model capture or address goal conflicts?
- does the model capture how an organization can manage or modulate its adaptive capacities (can an organization reflect upon its adaptive capacities)?
- does the model include how systems can restructure themselves as demands grow or persist?
- does the model include calibration or miscalibration – how well does the system understand or estimate its own adaptive capacities?
- does the model capture an optimality-brittleness tradeoff?

## Model Comparisons

Table 4.2 provides a preliminary comparison across four models of resilient systems. Looking over the table allows one to make some generalizations.

**Table 4.2      Comparison of concepts across models**

| Concept | Ecological Systems Theory (Holling, 1973) | SES (Walker et al., 2004) | Ball and Cup (Scheffer et al., 1993) | Stress-Strain (Woods and Wreathall, 2008) |
|---|---|---|---|---|
| Modes | Equilibria | Systems | Basins | Regions |
| Stability | Engineering resilience | Resistance | Cup slope | Young's Modulus |
| Amount of stress before restructuring | Ecological Resilience | Latitude | Cup height | Tensile strength |
| Precariousness | – | Yes | Yes | Yes |
| Reserves | Spatial mosaics | Passive | Increase basin height | Active |
| Decompensation pattern | – | – | – | Fundamental |
| Second-order adaptive capacity | – | – | – | Extra region; transitions over regions |
| Graceful degradation | – | – | – | Graceful degradation margin |
| Goal conflicts | – | – | – | Yes |
| Manage system adaptive capacities | – | Adaptability | – | Monitor operating point and trends in state space |
| Restructuring | – | Transformability | Multiple basins | Restructured region |
| (Mis)Calibration | – | – | – | Systematic monitoring failures |
| Optimality-brittleness tradeoff | – | – | – | Represented |

First, all of these four models provide some kind of fitness space representation where fitness is a relationship between a system's behavioural capacity and characteristics of the environment (e.g., variability, disrupting events, classes of demands). Models of adaptive systems require a means to specify what fitness is, how behaviour changes to achieve better match to demands, and how changes in the environment challenge the fitness of the behavioural repertoire previously developed. The question then focuses on what concepts about system resilience is highlighted or captured naturally in each fitness representation.

Second, Holling's original formulation and the ball and cup models primarily specify parameters that capture the base, or first-order, adaptive capacity of a system to handle new or changing demands. These parameters are also present in the Stress-Strain State-Space diagram.

Looking at the ball and cup metaphor from the side, we can see the similarity to the Stress-Strain diagram (truncate the left half of the left basin). The similarity runs deeper than visual appearance. The stability, or engineering resilience, in both cases refers to the slope of the uniform region in the Stress-Strain diagram (Young's Modulus in materials science) and represents the rigidity or flexibility to specific pressures. The ecological resilience (pressure required to force a transition) in both cases refers to the height of the curve. In materials science, this point is known as the ultimate yield strength.

Third, Social Ecological Systems and the Stress-Strain State Space both include the concept that human systems need to be able to monitor and manage their adaptive capacities. This is an important conclusion that was developed independently from the work on Social Ecological Systems (e.g., Walker et al., 2004; Folke et al., 2005) and from the work on resilience as a means for proactive safety management in complex systems. The latter motivated the need for models of system resilience such as the Safe Operating Envelope and the Stress-Strain State Space (e.g., Hollnagel, Woods, and Leveson, 2006; Woods, forthcoming). This ability to reflect about one's own adaptive capacity is uniquely associated with human systems.

Reflecting on the adaptive capacity of a system in which one participates as an operator or manager necessarily leads to the critical concept of calibration. Knowledge calibration refers to how well one knows what one knows. In other words, how accurate is one's knowledge of where the system is operating in its fitness space and what trends are developing? The Stress-Strain State

Space was explicitly developed to plot calibration/miscalibration (Woods and Wreathall, 2008). However, in principle, any fitness space representation can be used in the same way to plot management's or other distant parties' mental models and to compare them to independent estimates (Dekker, 2006).

Fourth, the Stress-Strain State Space naturally specifies several important resilience concepts that are missing from or difficult to represent in the other models:

- the transition from base adaptive capacity to additional or second-order forms of adaptive capacity – how a system adapts when events or change challenge its basic ways of functioning (Woods, 2006; Carlson and Doyle, 2002; Zhou et al., 2005);
- the decompensation pattern which is the basic behavioural pattern and failure form for adaptive systems (Woods and Cook, 2006);
- the graceful degradation margin (GDM) which is a potentially important measure of the resilience or brittleness of organizations;
- active deployable reserves are needed for resilience in human systems and, therefore, support for basic cognitive system functions such as anticipation, anomaly response, re-planning, synchronizing joint activities. Acting only to preserve passive reserves is very difficult to sustain in the face of pressures to achieve acute goals (Bengtsson et al., 2003; Woods, 2006);
- none of the representations directly handle how resilience arises from the need to balance goal tradeoffs (e.g., efficiency-safety) or more fundamental tradeoffs (acute-chronic perspectives).

Stepping back from the specifics of the comparison matrix, one can get a sense of the basic ways in which the label resilience has been used. One could group definitions of resilience into four general classes that are used to refer to:

- the parameters of a system's base adaptive capacity – what that system is designed or inherently can adapt to in terms of variations, factors, situations and change in the world;
- second-order forms of adaptive capacity – how the system adapts when events and change challenge its usual capacity to adapt;

- a general framework or space to represent different forms and aspects of a system's adaptive capacity;
- a system's ability to recognize and manage transitions across types of adaptive capacity – managing and modulating a system's varieties of adaptive capacity. The last is particularly interesting because two ideas come to the fore: (a) a resilient system is able to reflect about its own adaptive capacities and (b) resilience refers to the potential for future adaptive action.

As potential, adaptive capacity exists before disrupting events call upon that capacity, but one can assess adaptive capacity only through its exercise in the anticipation and response to past disruptions. This means that the resources (cognitive, collaborative, and physical) that support this potential may not be seen since they are not used prior to visible disrupting events, or, if seen, they will be interpreted as excess to be eliminated for efficiency goals. As a result, it is easy for an organization to be miscalibrated about its own adaptive capacities and the changing demands it faces.

## Highly Optimized Tolerance

In materials, strength and brittleness interact in many ways. The analogy to materials science raises an important question for organizations and social-ecological systems. Do systems become more brittle overall, even as they become more optimized to some set of demands and variations in the environment?

Again, biological systems research provides insights about general principles of adaptive systems. Carlson and Doyle (1999, 2002) introduced the concept of Highly Optimized Tolerance (HOT) to describe regularities in adaptive systems. They describe a basic tradeoff: as a system becomes better adapted to handle common or anticipated demands, it becomes increasingly fragile to rare or unanticipated demands, events, or design flaws. For example, those species that specialize more rapidly will proliferate, but are vulnerable to sudden changes in the environment. Those that remain generalists will be less efficient competitors, but will survive and proliferate during and after sudden changes in the environment (see also Stromberg and Carlson, 2006; Zhou, et al., 2005).

This fundamental tradeoff between optimality and resilience is a hallmark of the study of adaptive systems. Ultimately, the optimality-brittleness tradeoff in the HOT analyses plays out in two basic tradeoffs for all adaptive systems: (1) specialist-generalist and (2) acute-chronic (Zhou et al., 2005; Stromberg and Carlson, 2006; Woods, 2006). The former emphasizes the system's behavioural capacities – adapting to take advantage of learning leads to growth of specialist roles. But specialization increases vulnerability to some adaptive traps, for example, specialization leads to high performance silos with fragmented responses to issues that cross specializations. The latter emphasizes environmental demands – adapting to short-term pressures advances acute goals. But this pursuit of "faster, cheaper, better" undermines an organization's ability to recognize signs that chronic goals are at increasing risk – leading organizations to sacrifice chronic goals inadvertently (Weick et al., 1999; Woods, 2005, 2006, forthcoming).

These aspects of the optimality-brittleness tradeoff point out how the tradeoff is a kind of "no free lunch" principle, as in micro-economics and complexity theory (Wolpert and Macready, 1997). Increasing adaptation to some aspects of, and variations in, the environment must make that system less adapted to other aspects of and variations in the environment. This means the optimality-brittleness tradeoff requires the distinction between base (first-order) adaptive capacity (what the system has and is learning to be better adapted to) and second-order adaptive capacity (how the system responds when confronted with demands that fall outside of this competence envelope). All models or representations of system resilience need to include some means to capture this tradeoff.

The Stress-Strain State Space representation addresses the optimality-brittleness tradeoff in terms of state space transitions – how does the system move from uniform to extra region or sub-regions and then shift to a restructured form of operation? – and in terms of dynamics – how does improving system performance in the uniform region affect the system ability to adapt in the extra region? When organizations mistake reserves that support second-order adaptations for mere inefficiencies and reduce or eliminate those reserves, the organization is miscalibrated and acts in ways that undermine chronic goals such as safety.

The question then becomes: are systems trapped by the optimality-brittleness tradeoff to endure cycles of learning, specialization, change, and collapse (e.g., Holling et al., 2002)? For human systems, can stakeholders reflect about, monitor, manage, and modulate forms of adaptive capacity in the systems they hold a stake in? In other words, can people in various roles and at various levels in a complex adaptive system learn better ways to balance the tradeoffs inherent in adaptive systems in uncertain, changing, finite resource environments to avoid maladaptive traps?

The goal of engineering resilient systems is to find ways that better balance acute short-term performance (such as efficiency measures) with chronic long-term performance (such as safety) in potentially changing environments (Hollnagel, Woods, and Leveson, 2006; Folke et al., 2005; Anderies et al., 2004). Biology demonstrates that there are designs that can relax the tradeoff between being more optimal but brittle, or being inefficient but highly adaptable to new challenges. For example, species will attempt to optimize performance through phenotype convergence while maintaining genotype diversity to protect the species from sudden environmental changes. Complexity theory searches for architectures that better manage the tradeoffs (Zhou et al., 2002). Multi-agent simulations explore the factors that produce balanced co-adaptive behaviours and avoid maladaptive traps (Hong and Page, 2002; Roth, 2008). Proactive safety management helps organizations look ahead to notice the signs that risks are changing or increasing despite past records of success and increasing pressures for short-term performance (Hollnagel and Sundstrom, 2006).

## Conclusions

This model comparison is limited and preliminary. The authors bring only one perspective, and others can adjust and revise the comparison matrix beyond this starting point. Nevertheless, this comparison matrix is valuable.

First, it provides a starting point for model comparisons. The concepts used to contrast models can be extended to include new insights (e.g., how models address cross-scale interactions, adaptive dynamics and cycles). The comparison can be extended to include other models such as Hollnagel's (2004) functional resonance accident model (or see Marais and Saleh, 2008). The matrix can also serve as a starting point for dialog across research teams. This dialog can clarify what concepts are unique facets, and what concepts reinforce or extend core ideas. It can also crystallize the key differences across different theoretical and empirical stances.

Second, the comparison highlights that there are key concepts or foundational ideas behind the label "resilience," despite the multiple metaphors, labels and definitions. While researchers and modellers are likely to continue to argue about specifics, the comparison generally revealed a basic technical core.

Third, the comparison shows that attempts to settle on a single all-encompassing definition of resilience are unlikely to be useful. The study of how systems are resilient/brittle does not reveal a single property or parameter that is "resilience." Rather, as our model comparison reveals, there are multiple aspects to how a system adapts given different events, variations and changes in the environment.

Fourth, representations of a fitness space provide a means to chart the different relationships that relate to a system's adaptive capacities. Plotting changes and contrasts (e.g., calibration) provides feedback on the system's adaptive capacities and how these are changing due to internal and external adaptive forces.

Fifth, there can be advantages to using a single encompassing framework such as the Stress-Strain State Space. A single framework can eliminate confusion over the multiple concepts that have been associated with increasingly popular use of the label "resilience." The Stress-Strain State Space provides a broad set of landmarks, measures, and distinctive patterns to plot and summarize results from studies of how organizations are resilient and brittle.

Finally, any debate over the advantages of one or another representation pales in the face of the critical next step for progress in resilience engineering. We need to find and demonstrate ways to estimate the parameters of one or another of these or other models of resilience. When these are relevant and practical, actual organizations will be able to monitor their adaptive capacities and will be able to adjust them when signs of brittleness indicate those capacities are either eroding or ill-matched to the demands that are about to occur (Wreathall, 2006).

# Chapter 5

# Measuring Resilience

John Wreathall

Bistromathics itself is simply a revolutionary new way of understanding the behavior of numbers. Just as Einstein observed that space was not an absolute but depended on the observer's movement in space, and that time was not an absolute, but depended on the observer's movement in time, so it is now realized that numbers are not absolute, but depend on the observer's movement in restaurants.

*Douglas Noel Adams (1952–)*
*English humorist and science-fiction novelist*
Life, the Universe and Everything. *New York: Harmony Books, 1982.*

## Introduction

As Hollnagel has observed elsewhere in this volume, there are four related capabilities that characterize a resilient organization or system. One of these is the ability to be critical; that is, the ability to monitor the performance of items that may threaten its continued existence, including its own performance. While most (if not all) organizations that survive in the real world are preoccupied with measurements, from profitability and financial performance to safety and quality measures, the needs of measurements for resilience engineering need to be specified in terms of the business processes related to resilience.

Using a basic principle from systems engineering, that measurements should be based on an understanding of the underlying processes, we therefore need to have a means of describing these processes by means of one or more models of the organization. An early start for proposing a relevant perspective of what a suitable model should describe was made by Wreathall

(1989), though in a different context at a much earlier time; this model is shown in Figure 5.1.

Since the purpose of this figure is to illustrate a general concept, its detailed contents will not be discussed here. However it provides a useful perspective on what a more directly helpful model will provide. As can be seen, it is hypothesized that at the general level similar *functions* are performed at multiple levels of organizations (and beyond) to ensure that risk is managed effectively throughout the organization.

From the individual worker through to the entire enterprise, activities go on that generate risk (in the broadest sense, whether it be to worker safety, public safety, the environment or even to the economic survival of the enterprise). The risks generated by these activities need to be assessed and managed to maintain control of the risk levels within tolerable bounds. While the functions of risk making, risk assessing and risk managing are the same at each level of the enterprise, the means by which they are accomplished are very different. For instance, consider the case of an individual in a workgroup such as maintenance. The *risk making* to worker or public safety can involve working on equipment that has hazards associated with it, such as highly toxic components or high levels of radiation; risk making involves having the potential to release these hazards. Other hazards to the organization may include business or financial hazards – consider the impact of trader-related problems in arbitrage trades at Baring Bank in 1995 and at

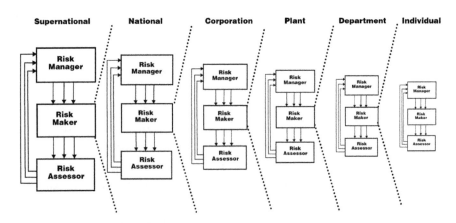

**Figure 5.1    Wreathall's original "Telescope" model**

Société Générale in 2007–2008, and the business decisions made by Northern Rock and the UK government following the run on its deposits.[1] Workers involved with such tasks would typically be well educated in the hazards and be trained in the tasks they are to perform. Therefore the individual worker can in many cases assess the levels of risk involved in performing the work (*risk assessing*) and make appropriate adjustments in the way they perform the tasks to ensure risk levels are maintained at a tolerable level (*risk managing*) – this is all done by the individual in an integrated and often not explicit manner. However, at a plant or corporate level, the production units will comprise the risk makers and formal parts of the organization will be performing risk-assessing activities (e.g., groups responsible for environmental or worker safety or quality), and the corporate or plant management is responsible for requiring changes in operations to meet risk levels set using tools such as probabilistic risk assessment (PRA).

Even broader than the levels of a commercial enterprise are activities belonging typically to nations, where industries are the risk makers, regulatory agencies such as Occupational Safety & Health Administration (OSHA) or Environmental Protection Agency (EPA), in the USA, perform risk-assessing activities and the legal or legislative systems direct the risk management activities. There can be supra-national bodies for some countries (such as the European Commission) or some industries (such as the International Atomic Energy Agency) that continue these functions at even more distant levels from the individual workers.

While the above model refers to risk making, etc., it is strongly relevant to resilience management. Resilience engineering (or resilience management) has been described as "managing the risks beyond risk assessment" (Wreathall, 2007). Thus resilience can be considered to be related to how well an enterprise is accomplishing, at all levels within itself, the activities represented in Figure 5.1. Therefore, included in the scope of resilience engineering are the following desires:

---

1    See, for example, the information presented on Wikipedia at http://en.wikipedia.org/wiki/Northern_Rock.

- The overall risk-management processes should include an adequate specification of the different types of risks that may threaten the enterprise:
  - worker safety;
  - infrastructure integrity;
  - environmental and public safety;
  - financial and business survival.
- The overall risk-management processes should be reviewed and updated at a frequency such that changes in scope and performance are detected and accommodated sufficiently often that the residual levels of risk are acceptable.
- The overall risk-management system should anticipate potential future hazards and conditions and take action to prevent being harmed significantly by them.

The challenge for finding measures of resilience is to develop measures inferring the extent to which these desires are accomplished. In order to do so, a modelling approach is proposed that will identify salient features of the overall risk-management process for which metrics can be identified.

### Development of Modelling

The representation in Figure 5.1 is simply to identify the necessary functions of risk management, and resilience is concerned with how well these functions are being performed. In order to make progress, what is needed is to develop one or more models to describe *how* these functions can be accomplished.

The underlying model in Figure 5.1 is recursive – that is, at the level of the functions to be accomplished, they repeat themselves at each level of the model, from the level of the individual to the highest level. However, these functions are implemented in different ways at the different levels. In order to explore the underlying processes, we first need a model that can describe in more detail the functions identified in Figure 5.1, and then we need a possibly different model that describes the means by which these functions are accomplished at the different levels. Two modelling approaches are proposed: first the viable systems model of Beer (1985) provides a recursive description of the processes to accomplish the functions in the Telescope model, and the soft systems modelling of Checkland (1981) and Checkland and Scholes (1990) allows us to develop customized descriptions of the processes within specific levels and technical domains.

*Viable Systems Model*

The viable systems model (VSM) provides a way of viewing those organizational processes that are necessary for an enterprise (the primary focus of Beer when considering "systems") to maintain independent survival in an integrated manner. It provides a recursive view in that the system has subsystems (and is, in turn, often part of a larger system), wherein similar processes exist. The origins of the VSM lie in the field of cybernetics, which seeks to describe and model self-governing systems. It represents the study of control and communications in the animal and the machine (to borrow from the title of Wiener's [1961] originating book). It is particularly salient here since the functions in the Telescope model are entirely concerned with issues of control and communication. Figure 5.2 illustrates the simplest representation for any one level in the Telescope model. Some *operations* take place that have products going to its *environment*, from whence it receives inputs. The operations are overseen by *management* whose activities provide directions as to modes of operations and which gather data about the performance of operations.

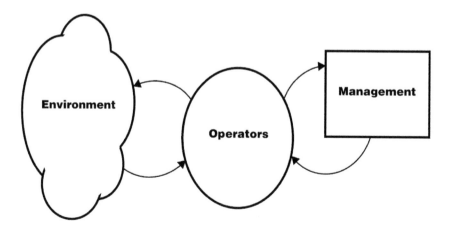

**Figure 5.2    Basic structure of system**

Thus, at the level of the individual in the Telescope model, the operations are performing the maintenance tasks on the equipment (for example) – they are transforming components that have failed into functioning equipment.

The management is provided by the technician's own observations and decisions about acceptable standards. Not only does the environment in this case include the equipment under repair, it also includes the equipment used by the technician, the work orders for the tasks to be performed and the work team of which he or she is part. It will also include most of the recipients of work outcomes, such as waste produced by the operations, and the public environment that could be harmed by potential releases of hazardous materials. As the level of consideration progresses up the telescope, some of these become part of "operations" (e.g., the equipment becomes part of operations when considering the plant).

This is a short summary of the VSM to provide a context for the remainder of the chapter; readers looking for a detailed description of the model itself are directed to Beer's work (1981, 1985) for more information. The VSM is concerned with the communications and control processes within the system represented in Figure 5.2, particularly as they relate to the control of complexity. In order to specify the particular activities, the VSM is considered to be composed of five different "systems" by which the overall goal of the system (self-sustaining operations) is accomplished. In the arcane language of the VSM, these five systems are simply referred to as "System 1," "System 2," and so on, which provides no description as to their locus or function. These will be briefly described below in order to explain their purpose related to the goal. Note that this description of the VSM is very much limited to the features relevant to the Telescope model. There are many features of the VSM that are overlooked, most notably its representation of the management of variety in the system – a key feature of the VSM in its intention.

System 1 represents the collection of operations activities that go on in parallel with one another, each interacting with its own local environment and under the influence of its own management. Thus, for maintenance at a plant, the scope of System 1 could be the assemblage of mechanical maintenance, electrical maintenance, instrumentation and control maintenance, and so on. These groups have separate functions and each has its own local management, but neither the groups nor their management operate in isolation from each other. Figure 5.3 illustrates the concept of System 1.

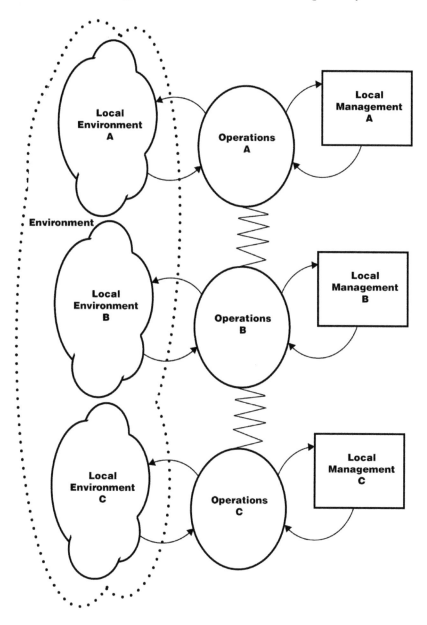

**Figure 5.3    System 1 of the VSM**

System 2 provides the means of reducing variability between the various components of System 1. This is illustrated in Figure 5.4 by the presence of the "dash-dot" lines from each of the local managements in System 1 to the "senior management" (a relative term). Examples of the mechanisms implicit in System 2 would be the use of standardized work processes, the safety climate, and

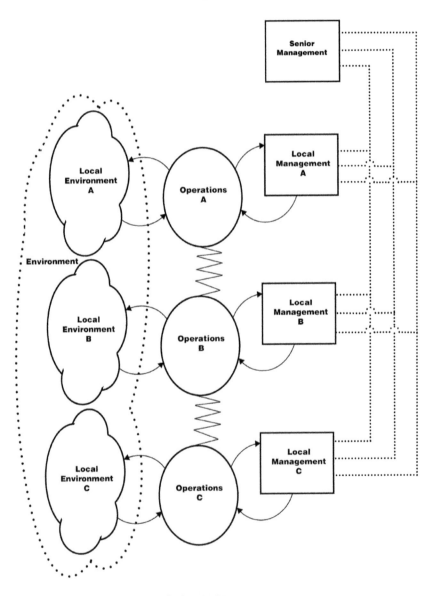

**Figure 5.4    System 2 of the VSM**

the work unit culture generally – these all lead to each System 1 group doing things in a similar way. Note that this is separate from top-down rules and regulations – the local culture may lead to these being obeyed or not.

Rather, the mechanisms in System 2 represent a local (often unstated) agreement between the local and senior management levels as to how work is done in practice – a kind of shared awareness. System 3 of the VSM is associated with supervision by the senior management of the operations activities, such as day-to-day control and auditing activities.

Within the construct of the VSM, the day-to-day management activities and periodic auditing activities are differentiated by referring to them as System 3 and System 3* respectively. Note that because of the growing complexity in Figure 5.5, only one pathway associated with this system is shown whereas in practice there would be multiple similar pathways associated with different kinds of audits, such as safety, business practices, finances, and so on. Not all operations would have all audit paths attached to them. For example, a group responsible for maintenance of non-radioactive equipment in a nuclear power plant would not have auditing associated with risks from nuclear hazards.

The scope of the senior management in this context is not just the oversight and coordination of the operating groups through Systems 2, 3 and 3* however. There are additional activities within their scope. One such set of activities is to look to the environment to see what kinds of changes can be anticipated and what impact they may have on the system. This is one of the primary functions of System 4, as shown in Figure 5.6. Thus the "senior management" activities relevant to this system would generally include market research, department and corporate planning functions (depending on the level in Figure 5.1), and research and development work. In the recursive nature of the VSM, these kinds of activities become some of the operational functions (i.e., System 1) of the senior management when it is the system of focus for the VSM model, at which point its senior management (e.g., corporate or executive levels) have their own System 4 activities, perhaps exploring major international business trends, shifts in government policies, and so on.[2] These multiple functions of

---

2    Topically, consider the failures in the banking and credit-related in-

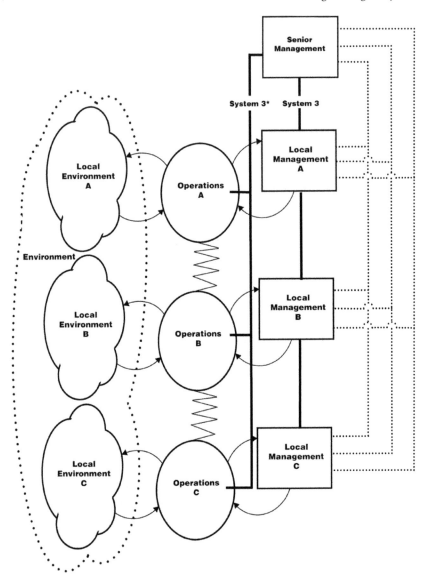

**Figure 5.5    System 3 of the VSM**

the senior management require that it maintain a balance of its activities, neither being given entirely to the internal relations with the operating groups nor entirely to the interactions with the environment.

dustries to anticipate the international impact of the so-called "sub-prime mortgage crisis" in the USA as failures in System 4 at almost all levels of the Telescope model.

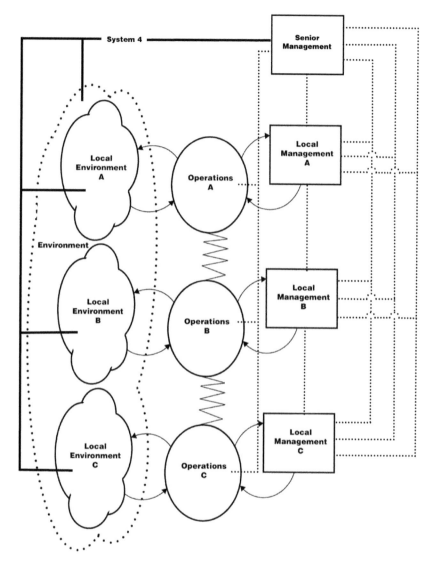

**Figure 5.6    System 4 of the VSM**

In addition to the anticipation of effects of the environment on the system, System 4 also provides a sense of self-awareness of the system itself – that is, it provides a sense of what the activities of the system are all about. This sense is what guides the senior management in terms of the relationships with the operating units through Systems 2 and 3.

The final part of the VSM, System 5, provides the next higher level of management functions in terms of setting an identity and

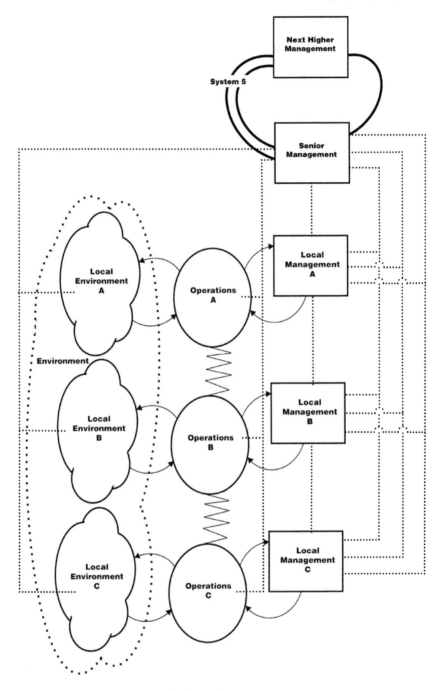

**Figure 5.7      System 5 of the VSM**

context for the work, monitoring the balance between Systems 3 and 4, and generally providing a basis for the overt value system within the VSM. In terms of the hierarchy presented in Figure 5.1, in many practical cases, Systems 4 and 5 represent the next higher levels of the organization. For instance, if System 1 were taken to represent individuals in (say) the maintenance department, then System 4 would represent the team of department heads and System 5 the plant management.

A critical aspect of resilient management is that it must have the ability to anticipate and prepare for potential future problems as well as learning appropriate lessons from accidents and incidents. Several functions within the VSM can be seen as being specifically useful for this purpose and are strongly relevant to the issue of resilience. First, at the level of System 1, the interactions with the local environments allows the operating entities to maintain an awareness of what is going on and what are the trends in these areas, *provided* the management at those levels are "tuned in" to look for the key signs of significant change. This perspective is usually provided by the senior management through System 3, which should also be maintaining its watching brief on the spectrum of environments through System 4. The motivation for the attention being given to the proactive assessments can be self-generated within System 1 (*bottom-up* proactive safety management, sometimes seen when one operating plant or division is spectacularly safer than its peers) or through the directives of senior and higher management via Systems 3 and 5 (*top-down* proactive safety management). (This is perhaps a good example of a limitation in using the VSM alone, because, while it talks about the *availability* and *effectiveness* of systems and structures, it says almost nothing about the *values* within which these operate. While values can be inferred from case studies for particular enterprises and activities – see the following example – it does not specify what these should be to achieve a particular goal, such as resilient or ultra-safe behavior (nor even satisficing[3] safety requirements "to make a buck").)

In order to illustrate an example of the VSM being applied in a real safety study, see the analysis of the 1999 Ladbroke Grove (London,

---

3      The principle that a satisfactory level or value is acceptable, even if more ideal solutions exist.

UK) rail crash by Santos-Reyes and Beard (2006) summarized in Table 5.1. The Ladbroke Grove accident (Cullen, 2000) involved the collision of a high-speed passenger train operated by First Great Western approaching Paddington Station and a passenger diesel-powered multiple unit (DMU) train operated by Thames Trains departing Paddington just after 8:00am on 5 October 1999, wherein the operator of the DMU passed a signal indicating a "stop" aspect. Thirty-two people were killed and 227 were hospitalized. The signal passed at danger (SPAD) was a known problem with a history of being passed at danger; no action had been taken to improve it. In addition the court of inquiry found numerous organizational contributions as summarized in Table 5.1. This analysis represents features of the British Rail operations and management following its privatization in 1994 that led to the separation of ownership of the rail tracks from the train operating companies when previously the entire rail system was united within British Rail. The following table summarizes how the components of the rail system correspond to Systems 1 to 5 in the analysis of Santos-Reyes and Beard.

*Soft Systems Methodology*

In the context of this work, soft systems methodology (SSM) refers to the methods and techniques that were largely developed initially by Prof. Checkland and others at the University of

**Table 5.1     A VSM analysis of the Ladbroke Grove accident**

| System | *Activities Relevant to UK Train System* |
|---|---|
| System 1 | Operating divisions *of the separate rail operating companies (specifically Thames Train and Great Western)* <br> Operations *by track operating company (Railtrack)* |
| System 2 | *Rail safety* coordination *between operating divisions and track companies – this was virtually absent in this event as a result of rail privatization* |
| System 3 | *Rail safety* functions *from the management of the operating companies to the operating divisions – again this was very weak as a result of the privatization* |
| System 3* | *Rail safety* auditing *within the companies – again an area of weakness* |
| System 4 | Developments *in rail safety – again an area of weakness, missing, for example, a growing trend in events involving "signals passed at danger" (SPADs)* |
| *System 5* | *Rail safety* policy *– again a weakness because of the privatization, there was no coordinating system to develop and manage policies* |

Lancaster (UK) starting in the late 1960s. It emerged from the tradition of systems engineering – the development of a model-based approach to understanding dynamic relationships in large engineered systems such as chemical process manufacturing. However, the adoption of the term "soft" indicated that the systems under consideration were importantly different from the so-called "hard" engineered systems (principally addressing plant equipment, physical processes, and so on) by focusing on "human activity systems" – that is, systems that are purposeful and whose characteristics are strongly influenced by the properties of being human: they are seemingly complex, full of ambiguity and very much driven by political (with a small "p") and social issues, as much as by business needs. Because of these characteristics, modelling techniques associated with hard systems engineering (rigorous definitions of process, fixed component boundaries, etc.) do not apply directly here.

In fact, SSM describes itself as an "enquiring process" in the sense of the process of uncovering issues and conflicts is a central part of the modelling process. Figure 5.8 (taken from Checkland and Scholes, 1990) shows the underlying model of the

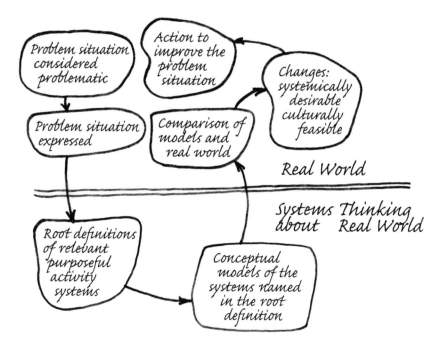

**Figure 5.8    Basic model of SSM**

SSM, which starts with some situation in the real world which is considered difficult – i.e., needing to be improved or resolved. The problematic condition is then expressed explicitly (since often difficulties in the real world are "felt" rather than explicitly stated unambiguously).

The systems approach then develops abstractions to better define and refine the nature of the problems through the use of "root definitions" and conceptual models of the systems relevant to the issue. The conceptual problems are refined so as to more completely describe the real world, and are then used to identify and suggest modifications in the real world that address the underlying problem within the context of the "human" dimensions of the system (i.e., within cultural and political constraints). An action plan is then created to implement the changes. In practice, further iterations take place to ensure that the changes are effective and that unanticipated consequences do not negate the intended benefits.

Within the SSM approach, then, the need exists to define the relevant systems and to characterize them through the creation of root definitions. The selection of which systems are relevant is somewhat subjective. In many cases there are generally two kinds of system that are relevant: primary-task systems that accomplish the business purposes, and issue-based systems that are relevant to the thought processes and inherent conflicts in the activities. In many cases the primary-task systems relate to formal business processes such as production, maintenance, operations and so on that are represented in the organization structure. Issue-based systems are much less obvious, rarely mapping to organizational charts. Rather than always being identified as distinctly separate systems, these terms relate to the extremes and most real-world systems have elements of both.

In identifying the relevant systems, the SSM approach is to identify them through the use of "root definitions." Root definitions in SSM are the labels that capture the essential core of the system and therefore require perhaps the greatest care in creating them. There are six characteristics to be defined in a root definition; these characteristics are known by the acronym, CATWOE:

- Customer(s): the organizations or people who benefit or suffer from the transformations accomplished by the system

- Actors: those who accomplish the transformations
- Transformation process(es): the conversions of inputs to outputs by the system
- World views (or *Weltenschlauugen*): the view(s) that makes the transformation meaningful in the context of this analysis
- Owner(s): those who can stop or substantially change the transformation
- Environment: elements outside the system that are taken as given.

The core components of the root definition are the pairing of the transformation and the world view. Checkland and Scholes (1990) provide a simple example for a public library of the different transformations according to two different world views:

Input→Transformation→Output

Local population→Transformation→Better informed local population

Books→Transformation→"Dog-eared" books

Clearly one of the world views relates to a societal function and one to a focus on the physical transformation. Once a root definition, based on these six elements, has been created, it provides the scope for the development of the system models. Typically for soft systems, the models will seek to describe the constituent practical (what is done), social (roles, values and norms) and political (power) systems embedded, since together they determine the effectiveness of the system as a whole.

Implicit in this approach is that systems again are recursive; systems contain systems and are, in turn, a part of larger systems. An example of a system description for an information system and then a system to measure its effectiveness is shown in Figure 5.9, taken from Checkland and Scholes (1990). Its root definition is:

A system, organized by the Information and Library Services Department [of a large corporate entity], which provides comprehensive information to active or passive users employing techniques and other skills, assisted by modern technology, so that the service is regarded as comprehensive.

The CATWOE mnemonic is interpreted here as:

- Customers: the users of the system
- Actors: ILSD staff, some (active) users
- Transformation process: users → users helped by provision of information
- World view: technology and the local (work) culture make this transformation feasible and useful
- Owner: the part of the company using ILSD as an agent
- Environment: organizational structure, current technology, company resources (assumed fixed within the time frame of the study).

Unlike the VSM, the SSM approach focuses on *how* things are done, not on what things have to be done.

**Means to Monitor Resilience**

The example system shown in Figure 5.9 is convenient in that it represents the development of a description to monitor the effectiveness of the information system in terms of efficiency and comprehensiveness. For the purposes of this chapter, we can consider these to be surrogate measures for resilience – see Wreathall (2006) concerning the identification of resilience-related metrics for comparison. It may be that resilience-related metrics also measure other aspects of the organization. In fact, I have found that such metrics already exist within organizations (for other normal business reasons) (Wreathall, 2001; Wreathall and Merritt, 2003), which is an advantage – the last thing any organization wants is to collect more data! What is important is that the lens through which they are seen brings them into perspective. The lens in this case is the set of relevant models of the organizations that describe the implicit processes introduced in Figure 5.1.

The VSM starts to provide an understanding of how the activities related to the monitoring functions take place. Consider, for example, Step 15 in Figure 5.9, "Define measures of performance for 'comprehensive'"; this would be accomplished principally through System 2 in the VSM – the system that creates

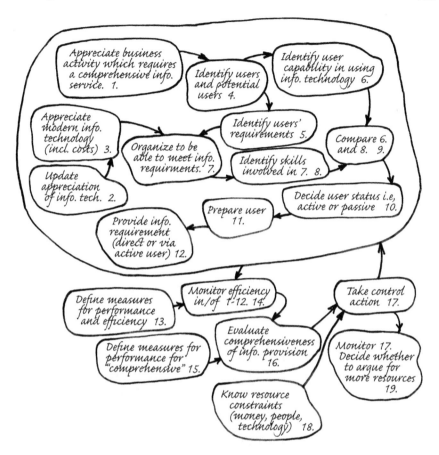

**Figure 5.9    Example SSM description for measurements of effectiveness**

a shared understanding of expectations of performance. Step 13, "Define measures of performance for efficiency," could be produced by a combination of System 2 (efficiency for the users, such as ease of access, perhaps) and System 5 when it comes to corporate concerns of efficiency (cost per user, etc.). Anticipation of future needs that could play a role in the planning of resources (Step 19) or the availability of new technologies that could permit the use of reduced resources (e.g., information access through personal data channels [e.g., Blackberry™ types of devices] rather than dedicated terminals) would be provided by System 4. In each of these cases, new SSM models could be created aimed at expanding the understanding of how these VSM systems accomplish their purposes.

Woods and Wreathall (2008) have identified one means to interpret resilience-related data through the use of a so-called stress-strain model, to provide a variety of signals concerning how close the organization is to failure, when there is a need for adaptation or reconfiguration to ensure survival of the system, and so on. Measures identified through the use of the combination of VSM and SSM modelling is well-suited to this approach. Work on combining these methods is continuing through collaboration with a large industrial organization to create specific measures to use in the management of losses – in this organization, "losses" include harm to people, the environment and the economic effects of performance failures, both to the organization and its customers. In other words, losses here represent those aspects of performance that resilience is intended to preserve.

## Conclusion

This chapter has laid out an approach that uses existing systems modelling methods (viable systems and soft systems modelling) in combination to create a means to identify the core processes (in terms of both *what* and *how* they are accomplished) related to the management of resilience, and how to suggest measurements associated with them. An example, based on published examples, is provided to illustrate the concepts. Work is continuing in collaboration with an industrial organization to implement these concepts.

# PART III
# Elements and Traits

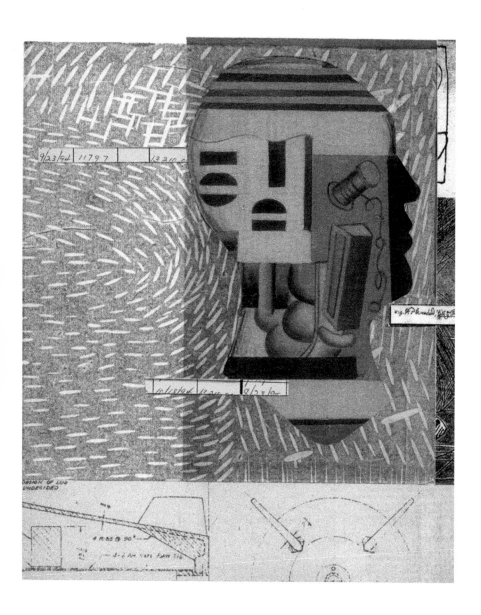

# Chapter 6

# The Four Cornerstones of Resilience Engineering

Erik Hollnagel

## Introduction

As the practical interest for resilience engineering continues to grow, so does the need for a clear definition and for practical methods. The purpose of this chapter is to propose a working definition of resilience and analyze it in some detail. The working definition is as follows:

> A resilient system is able effectively to adjust its functioning prior to, during, or following changes and disturbances, so that it can continue to perform as required after a disruption or a major mishap, and in the presence of continuous stresses.

The key term of this definition is the ability of a system to *adjust* its functioning. (The terms system and organization are used interchangeably in this chapter.) This makes clear that resilience is more than the ability to *continue* functioning in the presence of stress and disturbances. While the ability of a system or an organization to preserve and sustain its primary functions is important, this can be achieved by other and more traditional means. Continued functioning can, for instance, be achieved by isolating the system from the environment, or by making it impervious to exogenous disturbances. An example of that is the *defence-in-depth* principle, which means that there are multiple layers of barriers between the system and the environment in which it exists. The defence-in-depth solution can, of course, serve to protect either the system, the environment, or both. In the field of nuclear power generation, defence-in-depth is defined as:

> a hierarchical deployment of different levels of equipment and procedures in order to maintain the effectiveness of physical barriers placed between radioactive materials and workers, the public or the environment, in normal operation, anticipated operational occurrences and, for some barriers, in accidents at the plant. Defence in depth is implemented through design and operation to provide a graded protection against a wide variety of transients, incidents and accidents, including equipment failures and human errors within the plant and events initiated outside the plant

*(INSAG, 1995: 4).*

A simple example of guarding against external disturbances is when I shut the door to my office or study to prevent unwanted noises from being heard (a physical barrier system) as well as to indicate that disturbances are unwanted (a symbolic barrier system; cf. Hollnagel, 2004). A more complex example is the allocation of beds in a post-operative Intensive Care Unit (Cook, 2006). In the case of industrial production, the ability to continue functioning can be ensured by having sufficient buffers for the system, such as resources and supplies. A more extreme example is provided by central command and control facilities for military (and governments), which can either be airborne or located in well-protected centres on – or sometimes under – the ground.

To maintain functioning, despite external disturbances and disruptions, clearly has a cost, since it is expensive to build and maintain defences and inefficient to keep too many resources and supplies ready for deployment without actually using them. It is therefore not a universally applicable solution, and is sometimes sacrificed in the name of productivity improvements or cost reductions.

## First Considerations

As the key term emphasizes, the ability to go on functioning is a result of the ability to *adjust* the functioning, rather than to maintain it unchanged. This adjustment can in principle take place either after something has happened (be reactive) or before something happens (be proactive).

Reactive adjustment is by far the most common. For instance, if there is a major accident in a community, such as a large fire or an explosion, local hospitals will change their state of functioning to be prepared for a rush of people that have been hurt. (For a more extreme case see the description of the responses to a bus bombing by Cook and Nemeth, 2006.) Responding when something happens may, however, be insufficient to guarantee the system's safety and survivability. One reason is that a system can only be ready to respond to a limited set of events or conditions, either in the sense that it only recognizes a limited set of symptoms or in the sense that it only has the resources needed for certain kinds of events – and usually only for a limited duration. Vivid examples of that can be found in everyday events, the most conspicuous case in recent years being the (lack of) response by the Federal Emergence Management Agency (FEMA) to Hurricane Katrina in 2005 (e.g., Comfort and Haase, 2006). In the world of business, the failure of the Airbus company to recognize and effectively respond to the problems with the production of the A380 in June 2006, and the later failure of the Boeing company to do the same with the production of the 787 in September 2007, suggest that limited readiness is not an unusual phenomenon at all.

Going further into the proposed definition of resilience, a second key phrase is that the system must be able to adjust its functioning *prior to, during,* or *following* changes and disturbances. The ability to make adjustments prior to an event means that the system can change from a state of normal functioning to a state of heightened readiness *before* something happens. A state of readiness means that resources are allocated to match the needs of the expected event, that special functions are activated, and that defences are increased. A trivial example is to batten down the hatches to prepare for stormy weather, either literally or metaphorically. An everyday example from the world of aviation is to secure the seat belts before take-off and landing or during turbulence. In these cases the criteria for when to go from a normal state to a state of readiness are clear. In other cases it may be less obvious either because of a lack of experience or because the validity of indicators is questionable. An increased state of alertness should, of course, not last longer than necessary since it may consume resources that otherwise could be used for normal performance. An example of failing to re-adjust is the alert level for air travel in the US, which was raised from the yellow (elevated) level to the orange (high) level on 10 August 2006, and has remained there ever since (at least at the time of writing this, May 2008).

The ability to adjust *during* changes and disturbances, to respond when something happens, has already been mentioned and will be elaborated further below. The ability to adjust *following* changes and disturbances means that the experiences from events of the past are used to make decisions about structural or functional changes so that the system is better prepared for what may happen in the future. These changes are often focused on the causes, as determined by accident investigations, although such causes and explanations always must be seen relative to the accident models and the investigation methods that were used (Hollnagel, 2004 and 2008a).

The working definition of resilience can be made more detailed by noticing that it implies four cornerstones of resilience, each representing an essential system capability. The four cornerstones, or four essential capabilities, are (cf. Figure 6.1):

- knowing what to *do*, i.e., how to respond to regular and irregular disruptions and disturbances by adjusting normal functioning. This is the ability to address the *actual*;
- knowing what to *look for*, i.e., how to monitor that which is or could become a threat in the near term. The monitoring must cover both that which happens in the environment and that which happens in the system itself, i.e., its own performance. This is the ability to address the *critical*;
- knowing what to *expect*, i.e., how to anticipate developments and threats further into the future, such as potential disruptions, pressures, and their consequences. This is the ability to address the *potential*;
- knowing what *has happened*, i.e., how to learn from experience, in particular to learn the right lessons from the right experience. This is the ability to address the *factual*.

The rest of the chapter begins a more detailed description of each of the four cornerstones. This is followed by a brief proposal

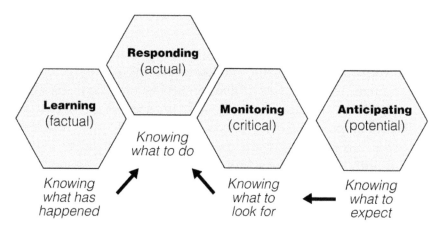

**Figure 6.1    The four cornerstones of resilience**

for practical ways in which this can be used to engineer the resilience of a system.

## The Actual

We will start by considering the *actual*, for the simple reason that this is the least that any system must be able to do. No system – be it an individual, a group, or an organization – can sustain its functioning and continue to exist unless it is able to respond to what happens. The response must furthermore be effective in the sense that it helps bring about a desired change. As described above, a resilient system responds by adjusting its functioning so that it better matches the new conditions. Other responses are to mitigate the effects of an adverse event, to prevent a further deterioration or spreading of effects, to restore the state that existed before the event or to resume the functioning that existed before, to change to standby conditions, and so on.

In order to respond when something happens the system must be able to *detect* that something has happened. Second, it must be able to *identify* the event and *recognize* or *rate* it as being so serious that a response is necessary. Third, the system must know how to *respond* and be capable of responding; in particular it must have or be able to command the required resources long enough for the response to have an effect.

The detection that something has happened is not entirely passive but depends on what the system looks for – on what its pre-defined categories of critical events or threats are. If the system looks for the wrong events or threats it may either fail to recognize some threats (false negatives or a Type II error) or respond to situations where a response was not actually needed (false positives or a Type I error). The former will leave the system vulnerable to unexpected events. The latter may be harmful both because the system may transition to a state that is not easily reversible, and because it wastes resources and reduces readiness.

Some events may be so obvious that they cannot be missed, yet without any response being ready – or even without a clear idea of what should be done. (The subprime crisis of 2007 was an example of that.) In such cases there may also be an urgency of the situation, i.e., an immediate pressure to act, which by its very nature limits the ability to consider what the proper response should be. A tragic example of that is the not infrequent situation when people get caught in a nightclub fire. Under such conditions the system, or in this case the individuals, may easily lose control by responding in an opportunistic or scrambled rather than in a more orderly mode (Hollnagel, 1998).

Rating or deciding whether an event or a threat is so serious that a response must be made can refer either to the establishing of a level of readiness, or to taking action in the concrete situation. In the first case, deciding that a response capability is needed depends on a number of factors: cultural, organizational, and situational. The dilemma is nicely captured by the common definition of safety as the freedom from unacceptable risks, which forces the question of when a risk is considered acceptable – and by whom. A common solution is to rely on probability calculations, and accept all risks where the probability is lower than some numerically defined limit (e.g., Amalberti, 2006). This, however, does not solve the problem of how the limit is set. Another solution is to invoke the *As Low As Reasonably Practicable* (ALARP) principle, which is defined as follows:

> A risk is ALARP if the cost of any reduction in that risk is grossly disproportionate to the benefit obtained from the reduction.

This links the acceptability of risks, and therefore also the decision of whether a response capability is necessary, to economical criteria. This becomes even more obvious in the UK Offshore Installations Regulations' clarification of the ALARP principle:

> If a measure is practicable and it cannot be shown that the cost of the measure is grossly disproportionate to the benefit gained, then the measure is considered reasonably practicable and should be implemented.

The second case is how to decide whether a response should be activated in a given situation. As long as the activation of the response depends on technology, including software, the problem is in principle solvable. But in cases where the decision depends on humans – at any level of an organization – the problem is more difficult. Deciding whether to do something, and when to do it, depends to a considerable extent on the competence of the people involved, and on the situation in which they find themselves (e.g., Dekker and Woods, 1999).

Finally, having the resources necessary for the chosen response is also essential. This is not only a question of having prepared resources, which really only makes sense for regular threats (cf. below), but also a question of whether the system is flexible enough to make the necessary resources available when needed.

The ability to address the actual can also be seen in relation to the various types of threats that may exist. Westrum (2006a) has proposed a characterization of threats in terms of their predictability, their potential to disrupt the system, and their origin (internal vs. external). He further proposed to make a distinction among three types of threats:

1. *Regular threats* that occur so often that it is both possible and cost-effective for the system to develop a standard response and to set resources aside for such situations.
2. *Irregular threats* or one-off events, for which it is virtually impossible to provide a standard response. The very number of such events also makes the cost of doing so prohibitive.
3. *Unexampled events*, which are so unexpected that they push the responders outside of their collective experience envelope. For

such events it is utterly impracticable to consider a prepared response, although their possible existence at least should be recognized.

The ability to address the actual depends on whether threats or events can be imagined, whether it is possible to prepare a response, and whether it is cost-effective to do so. Referring to the three categories listed above, it is clear that such readiness is only feasible for regular threats. However, that does not mean that irregular threats and unexampled events can be disregarded. They must, however, be dealt with in a different manner.

## The Critical

A resilient system must be able to flexibly monitor what is going on, including its own performance. The ability to monitor enables the system to cope with that which could become critical in the near term. The flexibility means that the monitoring basis must be assessed from time to time, so that the monitoring does not become constrained by routine and habits.

As argued above, it is in practice only possible for a system to be ready to respond to the regular threats, or even just to some of them. It is nevertheless a potential risk if the readiness to respond is limited to a small number of events or conditions. The solution is to monitor what may become critical, and use that to change from a state of normal operation to a state of readiness when the conditions indicate that a crisis, disturbances, or failure is imminent. Such a two-step approach will be more cost-effective. If a system can make itself ready when something is going to happen, rather than remain in a state of readiness more or less permanently, then resources may be freed for more productive purposes. The difficulty is, of course, in deciding that something may go wrong early enough so that there is sufficient time to change to a state of readiness. It is also necessary that the identification of the impending event is so reliable that preparations are not made in vain.

Monitoring normally looks for certain conditions or relies on certain indicators. These are by definition called leading indicators, because they indicate what may happen *before* it happens. Everyday life is replete with examples, such as the indicators for the weather tomorrow or for the coming winter (or summer). While meteorologists today are very good at predicting the weather, risks analysts are less successful in predicting when something will go wrong. (It is a sobering thought that economists and politicians seem to be even less capable of foreseeing financial and political crises.) In the case of the weather there are good leading indicators because we have an accurate understanding of the phenomenon, i.e., of how the (weather) system functions. In other cases, and particularly in safety related cases, we only have weak or incomplete descriptions of what goes on and therefore have no effective way of proposing or defining valid leading indicators. Because of this, most systems rely on lagging indicators instead, such as accident statistics. While many lagging indicators have a reasonable face validity, they are only known with a delay that often may be quite considerable (e.g., annual statistics). The dilemma of lagging indicators is that while the likelihood of success increases the smaller the lag is (because early interventions are more effective than late ones), the validity or certainty of the indicator increases the longer the lag (or sampling period) is.

According to the proposed definition, resilience is the ability of a system to effectively adjust its functioning prior to an event. As argued above, this can only be done if attention is paid to that which may become critical in the short term. For a system to do so requires that the effort is deemed worthwhile, that the necessary investment in resources (time and money) is made, and that the monitoring focuses on the right indicators or symptoms. Failing that, the system will sometimes be caught unprepared when it should have been ready. There will always be situations that completely defy both preparations and monitoring – the dreaded unexampled events – but more can be done to reduce their number and frequency of occurrence than established safety practices allow.

## The Potential

While looking for what may go wrong in the immediate future generally makes sense, it may be less obvious that there is an advantage to look very far ahead as well, to look at what could possibly happen in the more distant future. The difference between monitoring and looking ahead is not just that the time horizons are different (short versus long), but also that it is done in different ways. In monitoring, a set of pre-defined cues or indicators are checked to see if they change, and if it happens in a way that demands a readiness to respond. In looking for the potential, the goal is to identify possible future events, conditions, or state changes – internal or external to the system – that should be prevented or avoided. While monitoring tries to keep an eye on the regular threats, looking for the potential tries to identify the most likely irregular threats.

It may be argued with some justification that risk assessment already does look for the potential. But risk assessment is constrained because it relies on representations and methods that focus on linear combinations of discrete events, such as event trees and fault trees. Established risk assessment methods are developed for tractable systems where the principles of functioning are known, where descriptions do not contain too many details, where descriptions can be made relatively quickly, and where the system does not change while the description is being made (Hollnagel, 2008b). For such systems it may be acceptable to look for the failure potential in simple combinations of discrete events or linear extrapolations of the past. Many present-day systems of major interest for industrial safety are unfortunately intractable rather than tractable. This means that the principles of functioning are only partly or incompletely known, that the description is elaborate and contains many details, that it takes a long time to make, and that the system therefore changes while the description is made. In consequence of that there will never be a complete description of the system and it is therefore ill-advised to rely on established risk assessment methods.

Looking for the potential requires requisite imagination or the ability to imagine key aspects of the future (Westrum, 1993). As described by Adamski and Westrum (2003), requisite imagination is needed to know from which direction trouble is likely to arrive and to explore those factors that can affect outcomes in future contexts. The relevance of doing that is unfortunately not always obvious as the following example illustrates. In a recent discussion about whether it was safe to use meat from cloned animals for human consumption, the chief food expert of the US Food and Drug Administration claimed that it was beyond his imagination to even find a theory that would cause the food to be unsafe. This categorical rejection of the possibility that something could go wrong obviates any need to look to the potential, and perhaps also to monitor for the critical.

Even if the possibility that something could go wrong is acknowledged, thinking about the potential is fraught with difficulties. Many studies have, for instance, shown that human thinking makes use of a number of simplifying heuristics such as representativeness, recency, and anchoring (Tversky and Kahneman, 1974). While these may improve efficiency in normal working conditions, they severely restrict the more open-minded thinking that is necessary to look at the possible. Looking for the potential is also difficult because it requires a disciplined combination of individual or collective imagination. It can also be costly, both because it cannot be hurried but must take its time and because it deals with something that may happen so far into the future that benefits are rather uncertain. Relatively few organizations therefore allocate sufficient resources to look at the potential. However, a truly resilient organization realizes the need at least to do something.

## The Factual

A resilient system must be able to learn from experience. Although this is mentioned last, it is in many ways the basis for the ability to respond, to monitor, and to look ahead. (The four cornerstones are actually equally important, and it is only the limitations of a written text that forces one to be mentioned before the other.) To learn from experience sounds rather straightforward and few safety managers, administrators, or regulators will willingly disagree with that. Yet if it is to be done in an efficient and systematic manner, it requires careful planning and ample resources. The effectiveness of learning depends on what the basis for the learning is, i.e., which events or experiences are taken into account; on how the events are analyzed and understood; and on when and how often the learning takes place. This can be elaborated as follows (for a more detailed discussion, see Hollnagel, 2008a):

- Which events should be investigated and which should not? Since human, material, and temporal resources are always limited, it is necessary to separate the wheat from the chaff, to focus on what is important and to disregard what is unimportant. One common bias is to focus on failures and disregard successes, on the mistaken assumption that the two outcomes represent different underlying "processes." Investigations may further be limited to look only at events with serious outcomes (accidents) and disregard other adverse events such as incidents and unsafe acts. Another bias is to focus on adverse events that happen locally and to disregard experiences that could be learned from other places. Resilience engineering tries to overcome all of these biases.
- How should events be described? Anyone who has been involved in an accident investigation or risk assessment knows that there are no objective or true descriptions of events. The description depends on which data are collected, how they are coded or categorized, and not least how they are analyzed. The latter is perhaps the most important factor since the assumptions behind the chosen analysis method to a large degree determines the result (Hollnagel, 2004). In accident investigation, as in most other human endeavours, *What You Look For Is What You Find*. A root cause analysis will, for example, not give the same results as an epidemiological analysis, and the learning will therefore be different in the two cases.
- When and how should learning take place? This is primarily a question of whether learning should be discrete or continuous, i.e., whether should it be done whenever something has happened or on a more regular basis? If it only takes place after "important" events, then nothing is learned from "unimportant" events, which are by far the more frequent. If learning is more regular, then how often should it be done and how many resources should be allocated to do it?
- What should the locus of learning be, individual or organizational? In any given situation, performance is determined by a combination of three things. First, individual knowledge and

skills – as well as the individual's perception and assessment of the situation. Second, institutionalized knowledge, usually expressed by means of rules, regulations, procedures, policies, and norms. Third, the attitudes to which knowledge to use and how to behave, whether to comply with rules or rely on "common sense," whether to prioritize own achievements or the group, and so on. Learning from experience can be directed at any of these, but the "mechanisms" must be appropriate for the locus.

In learning from experience it is important to separate what is easy to learn from what is meaningful to learn. Experience is often couched in terms of the number or frequency of occurrence of some event or other, usually ones that are negative (accidents, incidents, loss time, etc.). But counting is not the same as learning. In order for a measure to be useful, it must be meaningful, hence referring to a principle, a model, or some kind of conceptual basis. While compiling extensive accident statistics may seem impressive it is not tantamount to learning and it does not mean that the system actually learns anything. Knowing how many accidents have occurred says nothing about why they have occurred, nor anything about the many situations when accidents did not occur. And without knowing *why* accidents occur, as well as knowing why they do *not* occur, it is impossible to propose effective ways to improve safety.

The difference between resilience engineering and the traditional view on safety can be summarized by looking at how the three questions above should be approached:

- A resilient system tries to understand how it functions, not just how it fails. Resilience is the ability to sustain normal functioning, not just to prevent failures. A resilient system should therefore not limit learning to specific categories of events and certainly not to failures rather than successes.
- A resilient system does not limit descriptions of events to their causes, as in the classical approach. Instead of looking for relations between causes and effects, resilience engineering looks for dependencies among functions and for the typical or representative variability of functions.

- In a resilient system, learning should be continuous rather than discrete, and should be driven by a plan or strategy rather than by events. One way of facilitating that is to try to learn from everyday situations and not just from situations where something has gone wrong. Indeed, if the focus is to learn from situations that turn out right, then learning will almost automatically become continuous. Learning from experience should also cover different spans of time. The lessons to be learned from short-term changes are not the same as the lessons to be learned from long-term trends, neither statistically nor causally. Finally, that which already has been learned should be revisited and revised because learning in itself will change the basis for learning and improve analysis methods. Lessons learned are never facts; they are interpretations that may have been valid when they were made, but where the validity is not guaranteed to last forever.

## How to Engineer Resilience?

If a resilient system is to be able to pay attention to the actual, the critical, the potential, and the factual, an obvious question is how this can be brought about. This is really the question of how resilience can be engineered or the question of what resilience engineering is in practice. While a detailed answer cannot be given here, a start will be made by considering each of the four cornerstones from a more operational perspective. This will give rise to a number of issues that in turn can serve as the starting point for more concrete measures.

- In order to address *the actual*, in order to know what to do, a system must have a clearly defined set of events to which it is ready to respond. So one step toward resilience is to develop or produce this set in a systematic manner. It is necessary to know why events are included in the set, both to be able to develop effective responses and in order to judge whether the events are still relevant. Failing to do that, the system may waste efforts in being prepared for events that should be of no concern, at the same time as it may be unable to respond to events that should be of concern. A second issue is how the readiness is established, i.e., how effective responses are formulated and how they are verified or ensured. A third issue is how the readiness is maintained. Keeping a system in a state of readiness is essential but costly. The resources needed may be manpower (e.g., staff on standby), knowledge (training and requalification), materials (energy, mass, and information), and so on. Maintaining readiness is not just a technical but also a management issue, and should therefore be addressed explicitly at the appropriate levels. The failure to verify and maintain an existing response capability is one of the ways in which latent conditions can arise.

- In order to address the *critical*, in order to know what to look for, the most important thing is a set of valid and reliable indicators. One question is how to define these indicators. The best solution is to base indicators on an articulated model of the critical processes of the system. Such models are, however, only feasible for pure technological systems. Another, more common, solution is to choose indicators that correspond to a tradition within an organization or a field of activity. A third solution, to choose indicators only because everyone else seems to use them (*cosi fan tutte*), should be avoided. A second question is how often this list is revised, and on what grounds. There should be clear guidelines for how to revise the indicators, how often, and on which basis. Quite often a revision takes place when something unexpected has happened, i.e., when the indicators have failed. But such a reaction is inappropriate both because it is unreasonable to expect the indicators to be complete and foretell everything, and because the revision will be hasty and superficial, putting more weight on face validity than content validity. A third question, already mentioned, is whether the indicators are leading or lagging, where leading indicators clearly are to be preferred. A fourth question is how measurements actually are made. Since most indicators will refer to some kind of aggregated measure, combining data sources of different quality, it is also important here that the rules and criteria are clear. A fifth question is whether the measurements refer to transient or stable changes, where the latter obviously is to be preferred. There should therefore be some way to determine whether a measured change is transient or stable.
- In looking at the *potential*, in looking towards the future, the most important issue is probably what the model of the future is. In

other words, what are the assumptions used to consider long-term developments? The simplest assumption is that the future will be a repetition of the past, i.e., that one should look out for a recurrence of past events, perhaps embellished with some degree of uncertainty. A less simple-minded assumption is that what can potentially happen can be found by an extrapolation of the past. This may even be developed into a formal model of the past that is used to calculate the future; most risk models fall within that category. A more realistic assumption is that what can potentially happen will be an emergent rather than a resultant phenomenon because the systems we deal with are only "nearly decomposable" (Simon, 1962). This means that every component of the system has a direct or indirect interaction with, in principle, every other component. Looking into the future must therefore be based on methods that go beyond cause-effect relations. A final issue is that looking at the potential in itself requires taking a risk, in the sense that it may lead to an investment in something that is not certain to happen. This again means that it is a management (blunt end) issue as much as an operational (sharp end) issue and that it should be treated as such.

- Finally, in looking at the *factual*, in trying to learn from the past, it is important to learn from successes as well as from failures. Successes are not just the near-misses or the spontaneous recoveries,

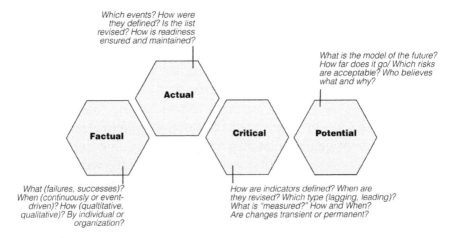

**Figure 6.2**     **Implementation     questions     for     resilience engineering**

but rather the normal functioning. If the focus is on accidents and adverse events, it means that only 1 out of 10,000 – or even 1 out of 100,000 – events will be considered. This is an enormous waste of opportunity unless, of course, it is assumed that successes and failures are the result of different underlying processes so that there is nothing to learn from the former. Resilience engineering strongly rejects this assumption. In consequence of that, leaning should be continuous rather than discrete. It is of course not feasible to try to learn from every event or every situation, both because it will impede productive work and because many of them are highly similar. But learning should follow a regular scheme rather than be part of the reaction when something goes wrong. Learning should also be both qualitative and quantitative. Quite apart from the fact that quantification is impossible without a prior qualitative analysis, more is usually learned by understanding what went on than by tallying specific outcome types, such as counting "human errors." A qualitative lesson will also be easier to communicate than numbers or statistics. Finally, learning should improve both individual and institutionalized knowledge.

The issues are summarized in Figure 6.2. The focus on the issues arising from each of the four cornerstones demonstrate how it is possible to think about resilience engineering in a practical manner. Starting from the level of the system as a whole this soon leads to the development of operational details and specific steps to be taken on a concrete level. This can, however, only be done by referring to a specific domain or field of activity, or even to a specific organization at a certain point in time. Much of that may obviously make use of existing methods and techniques, although seen from a resilience perspective and in some cases supplemented by new methods and techniques. For any given domain or organization it will also be necessary to determine the relative weight or importance of the four main abilities, i.e., how much of each is needed. The right proportion cannot be determined analytically, but must be based on expert knowledge of the system under considerations and with due consideration of the characteristics of the core business. Yet the minimum requirement is that none of them can be left out if a system wants to call itself resilient.

# Chapter 7

# Ready for Trouble: Two Faces of Resilience

Ron Westrum

To my mind, resilience has two faces. One is hardware, one is the human envelope that operates and protects the hardware. Each is important, but their logic is different. While much thought has been devoted to the former, building the latter must be seen as a key endeavour. My purpose here is to try to balance the picture by comparing the human envelope as a resilience mechanism to the use of requisite imagination in hardware design. We will use examples from NASA and other high-tech operations.

Let us contemplate two parts of NASA's *Apollo* moon-landing program. The first is the design of the *Lunar Lander*, a vehicle to put men directly on the moon (Kelly, 2001). This was an enormously complex design and fabrication process carried out by Grumman Aircraft Engineering Corporation on Long Island, New York. The task was to prepare a craft that could land on the lunar surface, disgorge two astronauts, and then re-launch to meet a moon-circling command module, which in turn would convey the astronauts back to earth. A daunting task: no one had been on the surface of the moon. And no one, it seemed, could be sure about all the things that might go wrong. Yet the equipment on the craft had to be ready for all the potential problems that could be thought up in the time allowed. One scenario after another was considered. Each had to have an answer, and that answer needed to be turned into hardware. The hardware had to work. Or the astronauts might die on the moon.

The second part to contemplate is the preparation of the flight control team for *Apollo 11*, the first of the *Apollo* missions to result in an actual moon landing.[1] This team was to be trained by Eugene Kranz, a very capable flight director. Unlike the *Lunar Lander*, the flight control team could not specify all the scenarios it might face. Instead, its strength would come from the depth of its knowledge, its flexibility, and its ability to fix something that had gone wrong. Exactly what would go wrong was not known. In preparation, the team worked through many exercises, making sure its cooperation was flawless. And in the process it needed to be familiar with every circuit diagram relevant to the operation of the spacecraft's many parts (Chaikin, 1995). In addition to the prime Flight Director, there were many accessory flight directors who all had their own backup. This preparation would mean the difference between success and failure when a major blowout took place on *Apollo 13* (Lovell and Kluger, 1995).

Resilience wears two faces. The first is physical and organizational: structures, procedures, and rules. The second is human and social: education, training, and practice. To put it simply: Do we put our trust in the people or the machines? Obviously, the answer in most cases is *both*. What I intend to do in this chapter, however, is to examine the logic of resilience in each of these two strategies, and show how these logics apply to actual situations. Finally we will note some differences in the strengths and weaknesses of these two ways of insuring resilience.

## Resilience in the Lunar Module

The moon lander program is a good place to start, because luckily we have a personal account by the chief engineer at the Grumman Corporation, the late Thomas J. Kelly (Kelly, 2001). One of the first tasks for Kelly and his team was to draw up a *Design Reference Mission* (DRM) document, which would define what Grumman and its contractors needed to design for. Kelly described the composition and functioning of the team to develop the DRM as follows:

---

1        Note that *Apollo 11* flight direction had two other highly capable teams, since it was a 24-hour operation. But it was Kranz's team that managed the actual moon landing, and it would be Kranz's team that managed the brainstorming for *Apollo 13*.

Thus was born the *Apollo* Mission Planning Task Force (AMPTF) in January 1964. With Barnes in charge, the AMPTF set up shop in one of the large *Apollo* conference rooms in plant 25 and was joined by team members from NAA, MIT, and NASA-Houston. Tom Barnes was a great team leader. Friendly, constructive and without ego or institutional bias, he inspired confidence and cooperation from the entire task force. Barnes was a talented systems engineer who explored problems relentlessly, asking key "what if" questions that sometimes led to new ways of defining or resolving things. He did it in such an easygoing but provocative manner that others were stimulated to new insights and contributions.

*(Kelly, 2001: 75)*

From this initial process and the resulting DRM the team defined what the spacecraft (and actually the mission itself) had to do, leading to a series of design studies. Eventually these led to a $1.4 billion contract to Grumman, for the building of 10 lunar landers (Kelly, 2001: 74). (Twenty were actually produced [McCurdy, 1993, 166].) We need not follow the rest of the engineering rollout in detail, but will concentrate on the process of designing equipment to meet the many scenarios for requirements and possibilities of mechanical failure. Through the engineering process the rough outlines of the sketches and prototypes were gradually evolved into drawings, and eventually into the hardware that became the landers.

Although this process is common to all engineering, we need to call out the process of requisite imagination, which Tony Adamski and I have called "the fine art of imagining what might go wrong" (Adamski and Westrum, 2003). The mission that the moon-lander team faced was extraordinary. Two astronauts had to descend to the surface of the moon from an orbiting spacecraft (the command module), land safely, and then get out and explore on foot. After the exploration, the two astronauts had to get back in the lander and take off, rejoin the orbiting command module, and head for the earth. If they couldn't take off, they would die. If they couldn't link up with the command module, they would die. It all had to function perfectly.

Everything had to be foreseen. Not just the ordinary steps of the mission, but every likely human error, every kind of equipment failure, every kind of environmental challenge. For instance, the nature of the lunar soil was known to vary, and the exact features of the landing zone could not be narrowed down. So the landing struts had to be able to cope with several possibilities. Even more significant was the recognition that the spacecraft system might need a lifeboat.

One major result of the AMPTF contingency planning was the identification of the "Lunar Model (LM) lifeboat" mission. While postulating the effect of various Command Space Module (CSM) failures on the outbound leg of the mission, the planners realized that a number of them could be countered by using the LM as a lifeboat and utilizing its propulsion, guidance and control, life support, and other systems to return the crew to the vicinity of the earth's atmosphere for re-entry in the CSM. To provide this rescue capability, some of the LM consumables, such as oxygen, water, and electrical power, would have to be increased by 10 to 15 per cent above that needed to perform the basic mission. Because LM then existed only on paper, we decided to make the tanks that much larger. At a later date it could be decided whether to actually load the additional consumables into them. Six years after it first appeared in the AMPTF's report, this vital mode was dramatically utilized on *Apollo 13* (Kelly, 2001: 76–77).

This use of requisite imagination would save the astronauts on *Apollo 13*, when the supply module suffered a major blowout. Although the mission itself could not be saved, the astronauts were returned safely to earth after they had stowed away in the lunar module. On nearing the earth, they climbed back into the command module and re-entered the atmosphere (Lovell and Kluger, 1995).

Again and again, the designers were forced to figure out what might go wrong and what to do about it.

Resilience in hardware, then, comes from building options into the hardware, and from minimizing those elements likely to bring the system down. The intelligent designer has the requisite imagination to understand the causes of failure, and to make sure that options exist to respond to them. At the same time he or she removes the latent pathogens that cut the margin of safety.

## The Erosion of Requisite Imagination in NASA

The explosions of the space shuttles Challenger (1986) and Columbia (2003) point to the long-term erosion of NASA's requisite imagination. It was not so much the destruction of the shuttles themselves, but rather the nature of the circumstances that allowed their destruction. Each demonstrated a failure of leadership. Each of these cases showed a failure to understand a basic flaw. And finally each demonstrated a failure to explore the dangers that faced the two Space Transportation Systems and respond to these dangers. While the failure of requisite imagination was not the only problem involved, it is worth seeing why requisite imagination eroded at NASA.

Requisite imagination has several preconditions. First of all it depends on expertise, that is to say, a fundamental understanding of the system and how it works. Part of expertise is learning from experience. In a narrow sense, experience means seeing what has happened before. But more broadly, experience means developing judgment about the kinds of things that are likely to go wrong, what can be trusted, and what cannot be trusted. In the context of the Columbia failure, the decline of a hands-on culture at NASA was remarked on both by astronomer James Van Allen and Daniel Baker, formerly of NASA-Goddard Space Center (Glanz et al., 2003). Both noted the absence of mature judgment in allowing the launch of Challenger. Would Wernher von Braun have run over the opinions of Thiokol engineers, or would he have demanded a more in-depth exploration? Similarly, in the case of Columbia, a more mature judgment would have pushed hard on the Crater model of foam damage used by Boeing. It certainly would have listened to the arguments and evidence brought forward by NASA structural engineers such as Rodney Rocha (Cabbage and Harwood, 2004: 125–45). Experience says you do not simply dismiss problems, you investigate until you are sure. It seems extraordinary today that the Air Force, quite willing to provide a photographic examination of Columbia in space, was told on two occasions that its additional information was not needed (Cabbage and Harwood, 2004: 111).

Requisite imagination also depends on the will to think. This phrase was used by Enrico Fermi to suggest that when one feels a problem can be solved, hope stimulates the mind to think about it (Shockley, 1974: 47). Clearly this was a major factor in the loss of Columbia. Many of the team felt that even if a flaw existed, nothing could be done about it. Ron Dittemore, the shuttle program manager, said: "We asked ourselves, is there any other option? There's no other option. If you want to come back home, have got to come back home through the atmosphere" (Cabbage and Harwood, 2004: 180). This fatalism forms a stark contrast with the "full court press" of *Apollo 13*, when not just team members but seemingly the whole world was asked about what to do. For Flight Director Eugene Kranz, "failure was not an option." He refused to give up hope. Requisite imagination might also mean the will to think about how to save an apparently "hopeless" situation, as it did on *Apollo 13*.

Finally, requisite imagination depends on inside knowledge. An expert on communication at NASA, Phillip Tompkins, in the Challenger explosion, saw a loss of inside information. In contrast, a key feature of NASA's Marshall Space Flight Center was that its managers had *penetration* into its contractor organizations during the *Apollo* era. For instance, when the Centaur rocket program was transferred to Marshall from the Air Force, the Air Force was using eight people to monitor the contractor. Marshall assigned 140! Marshall thus knew what was going on at the contractor organizations, what was done well and what wasn't. Contractors' first contact with Marshall was often traumatic, since not only did Marshall personnel ask more questions, they often had better answers to them than the contractor did (Tompkins, 2005: 89). To Tompkins the circumstances of the *Challenger* accident demonstrated that NASA had lost this *penetration*.

With the Shuttle program, all three of the critical preconditions had been eroded by poor leadership, budget cuts, and a flawed organizational culture, a decline from the *Apollo* years to the time the shuttles were launched. In a 1988 survey carried out by Howard McCurdy, this loss of in-house technical capability was marked, especially for employees hired from 1951–1969. Some 63 per cent of the 704 managers, engineers, and scientists surveyed thought that NASA's in-house technical capability had declined, and 78 per cent felt that too much of the agency's work had been turned over to contractors (McCurdy, 1993: 181). Of the earlier hires, in the same survey 81 per cent noted a marked decline in exceptional people recruited by NASA (McCurdy, 1993: 175). A part of this decline was a critical absence of technological maestros (see below).

I noted that a similar decline took place at the China Lake Naval Weapons Center after the 1960s, when budgets were cut and design work was transferred elsewhere (Westrum, 1999: 254–73). Requisite imagination fades when it is not tempered by the practice of engineering. When China Lake changed from an organization that designed weapons to one that managed a design process happening elsewhere, hands-on experience became thin. The paradox: "How can you manage design when you don't do it?" is all too real. Or as Earl Butz, a former US Secretary of Agriculture is reported to have said about the Pope ruling on premarital sex: "He no playa da game, he no maka da rules!" Without the experience, it is hard to have a "community of good judgment," where those in charge have "been there and done that" (Westrum, 1999: 247).

## Strength in People: The Human Envelope

While much study has been done on the process of engineering and project management (Morris, 1994)' much less has been done on the engineering of the human envelope that protects these systems (Westrum and Adamski, 1999). Around every sociotechnical system, there is a human envelope of care. This human envelope includes many different groups: the designers, the operators, the managers, technical support, the regulators, and so on. The social architecture and management of the human envelope is key to the success of large technological systems. Often the design and management of the human envelope is a key task undertaken by the project director, a *technological maestro*, in the phrase of Arthur Squires (Bowser, 1987).

A technological maestro, says Squires, has a number of key traits (Squires, 1986). The first is high energy, which allows the maestro to do two other things usually thought to be contradictory. One is to hold on the key ideas, and the other is to manage the details. Maestros can do both; they can operate at a high level of abstraction but also pay attention to the critical details. Also the maestro needs to be a technical virtuoso, so that he or she knows the actual technical issues well. A maestro needs to enforce high standards, so that the troops will be motivated to work up to them. A maestro needs to be hands-on, so that he or she has the *penetration* mentioned above. Squires has shown, in a variety of cases, how maestros were needed to make big technological projects work, from the Panama Canal to *Apollo* itself (Squires, 1986).

Two examples of technological maestros are Joseph Strauss, chief engineer of the Golden Gate Bridge (near San Francisco), and Wernher von Braun, rocket engineer and scientist, one of the guiding lights of early NASA. Strauss, in designing the Golden Gate Bridge (which opened in 1937), was concerned about loss of life in the construction of previous bridges (Coel, 1987). The rule of thumb at the time was that there was one life lost for every million dollars spent on building the bridge. Strauss was determined that this would not be true of the Golden Gate Bridge, so he took extra steps for safety. The first was to make his workforce wear glasses that would allow them to see through fog. The second was to get the men to wear hard hats, at that time an innovation. The third, and perhaps most impressive, was to stretch a hemp net from shore to shore under the bridge. These innovations proved their worth. With the hemp net, for instance, we know the value, because 19 men fell into it during the construction process. Later a travelling crane fell into it, with 10 men who lost their lives. The net parted, an unforeseen eventuality, underlining how hard it is to foresee everything.

Another expert in constructing the human envelope was Wernher von Braun, brilliant engineer and superb project manager (Neufeld, 2007). In von Braun's administration of the Redstone Arsenal, later the Marshall Space Flight Center, he de-emphasized hierarchy and encouraged open technical communication. By encouraging the Center's 7,000 employees to communicate freely and effectively, von Braun insured better design and manufacture of the many projects undertaken. The "bottle of champagne" given to an engineer who had admitted causing a prototype failure is now a well-known story (von Braun, 1956). But in fact von Braun's emphasis on honesty, and his discouragement of barriers to communication, was systematic (Tompkins, 1993).

So what does it require to build a human envelope? What does one have to do to make a sound and resilient human envelope? There are some preliminary requirements.

The first requirement is *training*. People must have the information to do their jobs correctly. Lack of training is associated with errors and accidents. The team that goes into action is shaped by the quality of its training. It is absolutely essential that the training be intense and realistic. It is vital for the operators to understand the strengths of the hardware, so that they can take full advantage of what is there.[2] The original function of the "Top Gun" training centre for Navy pilots – originally at NAS Miramar, now the Naval Strike and Air Warfare in Fallon, Nevada – was to train the pilots to use the capabilities of the F-4 Phantom fighter-bomber. The famous Ault Report, in 1969, had underlined the contrast between what the Phantom could do versus what the pilots knew how to use. Top Gun satisfied a key requirement of the Ault Report, the need for a "graduate school for fighter pilots." The hardware's strengths can only be used if the pilots understand them (Wilcox, 1990: 90).

The basic principle of the human envelope is to *engineer for capability*. Unlike the design of a spacecraft, the design of the human envelope does not try to envision everything that might happen. Rather, it tries to prepare a set of people to handle whatever it is that *does* happen. This is key, because experience has shown that – good as requisite imagination is – it is never perfect and one has to be ready to confront the unexpected.

For the *Apollo 11*, mission training was intense and profound. Eugene Kranz, NASA Flight Director, made sure that his team trained intensively for every scenario that they could think of. Everything was done to provide a readily-available store of knowledge that would save them in the result of an emergency. In the end the level of coordination was near perfect.

> Kranz knew his team of flight controllers the way a battle commander knows his troops. These young men, like all the people who worked for [flight director] Chris Kraft, were exceedingly bright. It was extraordinary what these people could do, together, during a mission. Kranz had seen it, time and again, in the

---

2    The China Lake Naval Weapons Center, during the Vietnam era, always made sure that the pilots understood the strengths and weaknesses of the weapons that China Lake had designed. Before deployment, the Commanders of the Air Groups (CAGs) would come out to China Lake and the weapons designers would talk to them. The designers also knew that there would be a debriefing when the air groups returned, and they would then be told what had worked and what hadn't (see Westrum, 1999: 224).

compressed moments of a launch or some other critical phase; they became one giant conglomerate brain, twenty minds wired together in parallel, each focused on some small piece of the whole event. In such situations, they could solve almost any problem that came their way, given twenty seconds to work on it. In twenty seconds a controller could look at the problem, talk to someone in his back room, think, talk to someone else, come back to that first person, and make a decision. And all the while, he would be monitoring events around him, listening not only to the conversations on the flight director's loop, but the air-to-ground, and perhaps one or two other loops. The amount of information possessed by one controller was staggering. And the entire team was trained for that kind of split-computer mentality. With that kind of brain power at work, twenty seconds could be a long time.

*(Chaikin, 1998: 171)*

This training of the group paid off. The extraordinary crisis of *Apollo 13* ("Houston, we have a problem!") and its resolution is well known. Less well known is what happened after *Apollo 12* suffered one, or perhaps two, lightning strikes shortly after launch. Many of the electrical systems, including the reference platform that oriented the spacecraft, suddenly went off-line. One of the assistant controllers, John Aaron, saw his own data displays disappear, then reappear in jumbled form. Remembering an experience during training, he spoke up strongly. "Flight, have the crew take the SCE to aux!" No one else seemed to understand what Aaron meant. He had to repeat himself slowly. But finally Alan Bean, on board the spacecraft, shifted the SCE switch to "aux" (power). This solved the problem, the data displays returned, and the mission could continue. Bean had worked on a similar scenario during training, and hence his confidence that this would solve the problem. But was anything else wrong with *Apollo 11*? Was the spacecraft still flight-worthy? In case there was any doubt, Chris Kraft and other flight directors encouraged the mission controller, Griffin, to speak out if there was any substantial doubt. There wasn't, and *Apollo 12* went to the moon (Kranz, 2000: 300–301).

Reflecting on the role of training in building resilience, William Tindall, another NASA maestro, said this about *Apollo*:

I think one of the greatest contributors to minimizing risk was the extraordinary amount of training that was done, high-fidelity simulations that were extraordinary. And there was no question about it, they saved us. I mean really saved us, many, many times because I don't think there was a single mission that we didn't have some significant failures. *The fact was that people could figure them out because they had been trained and knew how to work with each other.* The communications were there, the procedures were there to figure out what to do in real time and get the thing going. And most of the time when those things happened, the outside world didn't know about it.

*(Logsdon, 1999: 28; italics added)*

The second requirement is the responsibility of the technological maestro to create a circuit in which there is a free flow of information. Many technological systems have come to grief because someone failed to speak up. In the case of the Hyatt Regency walkway collapse, workmen had previously discovered the walkways were not sound. But no one had spoken up, probably because the workmen did not feel that they were part of the team (Petroski, 1992). Whether it is a question of removing the latent pathogen, or seeing a technical fix that no one else has yet seen, this free flow of information is essential. And people need to be trained to feel that they have the right to go forward with whatever their grasp of the situation can do to advance the system. Christopher Kraft observed about the *Apollo* program:

We all, there is not a soul here or was in our organizations that felt like they couldn't say what they wanted to say anytime they wanted to say it and felt totally comfortable about it. We grew, that was our heritage. It was the way we thought. We were never embarrassed about it. We were never embarrassed about being made a fool of when we made mistakes because we made them. I mean, we made hundreds of them. But we were used to being open about them. And that was fundamental to getting our job done.

*(Logsdon, 1999: 33)*

The willingness to communicate must be matched by the willingness to listen. An anecdote about William Franklin "Boss" Kettering (Boyd, 1957) relates a story in which he took seriously some remarks by a member of the organization who was not even in the chain of command. Kettering, who designed diesel engines, had come down to the dry docks to observe how his engines fit their ships. A painter was standing on a walkway staring at the stern of a ship. Kettering inquired what he was looking at. The propeller, the painter said, was too big, perhaps by five and a half inches. Kettering pulled up a nail keg, sat down, and talked more to the painter. In the end he called the design office and asked if they had measured the propeller. Of course, they said.

"Well, would you come down and measure it again?" They did, and found the blade was off by nearly the amount the painter had guessed (McGregor, 1966, 119). This ability to process "faint signals" can often be decisive.

The readiness to react creatively may be just as necessary with the maestro as with the team that he or she creates. When a student, doing a term paper, called William LeMessurier, an engineering consultant for the 59-story Citicorp building, he set off an important chain of events. The building's unusual footprint didn't seem stable to the student. How would it respond to, for instance, a quartering wind? LeMessurier assured him that it would do just that; it had been designed to handle all kinds of wind stresses. But then the engineer got to thinking, had everything gone according to plan? He called the contractor to make sure that there were no important changes between conception and execution. It turned out that changes had been made. Vertical beams had been secured with rivets instead of welds. This was a normal substitution, and normally it would not matter. But calculations showed that in this case it *would* matter, notably in the case of a severe quartering wind, one that would occur, on average, every 60 years. Quietly, changes were made to the building, already occupied. Every night after the secretaries went home, wall panels were removed and vertical beams were welded. Police and newspapers alike were notified but agreed to let the job proceed without public notice. When the job was finished, the public was let in on the secret (Morgenstern, 1995).

Third, it is the duty of the maestro to secure the loyalty of everyone in contact with the system, since it is never clear who will spot the latent pathogen or the potential improvement. It is never clear who will come up with the crucial piece of information. This building of the emotional bond of the network is absolutely critical. We have seen how the Hyatt Regency walkways could have been saved if the workmen had felt themselves to be part of the construction team. In the flooding following Hurricane Katrina in New Orleans, the official agencies involved mostly failed (Brinkley, 2006).[3] It was all the individuals and groups that had *not* been officially tasked with saving lives that largely

---

3       Two exceptions were the Louisiana Department of Fisheries and Wildlife, and the US Coast Guard (see Brinkley, 2006).

saved people who had been unable or unwilling to leave the city (Westrum, 2006b). It was their devotion to duty and charity that counted.

A counter-example, of failed loyalty, was the failure by managers of the *Columbia* space shuttle for its final and fatal trip in January 2003. When the shuttle launched, three pieces of foam hit the leading edge of the shuttle's left wing. This "foam strike" later resulted in the shuttle exploding on re-entry to the atmosphere. During the flight, however, those responsible for the shuttle's safety repeatedly brushed off indications that all was not well. Relying on a badly understood computer model, and some grainy photographs, the managers concluded that the foam strike did not pose a "safety-of-flight" issue (Wald and Schwartz, 2003). When structural engineers, for instance, tried to get some photographs from the Air Force, the Mission Management Team leader had the photographs cancelled, in spite of the willingness of the Air Force to provide them. Meetings of the Mission Management Team (MMT) were done briskly, and the "need for closure" – a symptom of groupthink – was all too evident. Although the MMT was supposed to meet every day, in fact it only had five meetings during the 16-day mission, and took off both weekends (one of them a three-day weekend) while the astronauts remained at risk (Cabbage and Harwood, 2004: 125–45). The structural engineers' concerns were brushed off not only by the MMT's leader, Linda Ham, but also by various other higher managers associated with the management of the shuttle. Those who needed to know did not get the information they needed, because they failed sufficiently to identify with those at risk and with the mission itself.

## Conclusion

In looking at the ways that requisite imagination was used to build resilience into the lunar module, we have seen the importance of making sure that everything likely to happen is covered. The technical options built into the lander protected the astronauts who landed on the moon and the crew of *Apollo 13*, who needed to survive a crisis in space. But at the same time, the options given to the spacecraft were only as good as the training given to the astronauts who would use it, and the training level of Flight Control, who would manage the missions from earth.

It is a prime duty of those who build systems, then, to make sure that the human envelope is given the same kind of attention given to the physical structure. The architecting of the human envelope is vital so that:

- those who are part of it can take advantage of the options that the technology presents;
- those in the system form the conglomerate brain that uses the information available to them in an efficient way;
- the system is surrounded by a *continuous web of thought* that constantly raises questions about whether the way the system is designed or functioning now is the way that it should be.

The human envelope is integral to the resilience of the system.

# Chapter 8

# Layered Resilience

Philip J. Smith
Amy L. Spencer
Charles E. Billings

## Introduction

Merriam Webster's *Collegiate Dictionary* (2003) defines resilience as:

> "1. The ability to recover quickly from illness, change or misfortune; buoyancy. 2. The property of a material that enables it to resume its original shape or position after being bent, stretched, or compressed; elasticity"

> *(p. 1060).*

The *Oxford English Dictionary*, Edition II, similarly defines resilience to be:

> "(1) Tendency to rebound or recoil; (2) Tendency to return to a state; ... (4) Buoyancy, power of recovery"

> *(p. 714).*

and resiliency as:

> "(1) rebound; recoil; (2) elasticity; the power of resuming the original shape or position after compression, bending"

> *(p. 714).*

Concern over the resilience of technological and sociotechnical systems (Hollnagel, 2006; Leveson et al., 2006; Weick and Sutcliffe,

2007) has existed for a long time. For instance, early in the history of aviation, Jackman et al. (1910) noted:

> One comparatively new design is the Van Anden biplane ... One flight, made October 19th, 1909, is of particular interest as showing the practicability of an automatic stabilizing device installed by the inventor. The machine was caught in a sudden severe gust of wind and keeled over, but almost immediately righted itself, thus demonstrating in a most satisfactory manner the value of one new attachment. ... It was the successful working of this device which righted the Van Anden machine when it was overturned in the squall of October 19th, 1909.

*(pp. 155–58)*

Below, we review a variety of different approaches to the design of a resilient human-machine system from a cognitive systems engineering perspective (Smith et al., 2008). Like the classic definitions of resilience provided above, these approaches emphasize "recovery." However, unlike some of these definitions, in some resilient systems this recovery does not necessarily return the system to its "original shape" or "state." Rather, the recovery can be a transition to some new, but acceptable or desirable state. In addition, the discussion below emphasizes different conceptual approaches for designing resilience into the system, including methods that rely on having an engaged human operator (or distributed team of operators) present to deal with unanticipated scenarios (Guerlain, et al., 1999), as well as methods that rely upon technology or, more accurately, that rely upon the ability of the designers to predict possible scenarios and to incorporate appropriate coping mechanisms into the technology (Leveson et al., 2006).

In general terms, then, this discussion illustrates different approaches to designing resilience into a human-machine system in which a person (or group of people) is supported by various types of technologies. It further discusses how hybrid combinations of these approaches provide a layered approach to resilience:

*Approach 1: Resilience Based on Detailed Predictions of Hazardous Scenarios or Causal Chains*

One approach involves resilience that has been designed into a system to deal with scenarios that have been explicitly predicted in detail, and for which explicit solutions (resilient responses) have been provided.

*Approach 2: Resilience Based on Specific Predictions of Hazardous "Effects" or "States"*

A second approach to designing resilience into a system involves developing the system so that it has the capacity to deal with a range of scenarios that may emerge and that have not been fully detailed in terms of "cause-effect" relationships, but that can be characterized in terms of some detectable "effect" or "state." Note that this distinction is an important one, as it implies that resilience can be introduced into a system without having to fully understand or predict all of the relevant causal chains.

*Approach 3: Resilience in Dealing with a Totally Unanticipated Scenario through Human-Centred Design*

One of the classic arguments for human-centred design (Billings, 1996; Smith et al., 1997) has been the contention that, when confronted with some novel situation, with access to sufficient information to detect and diagnose the situation, and with a sufficient range of possible responses, a person or a group of people can exhibit the flexibility and intelligence necessary to make an effective response in spite of the fact that a situation was totally unanticipated "due to an unexpected combination or aggregation of conditions or events" (Hollnagel, 2006: 12), even at an abstract level (Helmreich and Meritt, 1998; Perrow, 1984). Note that in this case, it could be that the state of the "world" has remained stationary, and that the system designer simply did not understand the complexities of that "world" sufficiently well to anticipate a particular scenario. However, it could also be that the state of the "world" itself changed in some fundamental way, such that a previously effective design is no longer sufficient. Note also that the claim that human intelligence is required is based on an assumption that the technologies underlying current approaches to artificial intelligence (or other forms of decision support) are either based on algorithms that the designer has explicitly programmed to deal with anticipated scenarios, or are based on learning algorithms that have made inferences from a number of previously observed cases.

*Approach 4: Layered Resilience or Hybrid Combinations of Approaches 1-4*

The three approaches outlined above are potentially complementary. In general, the designer should try to predict and plan for potential contingencies in order to ensure an effective, customized response to each such situation (Approach 1). However, since it is not always possible to accomplish this for all contingencies, a second line or layer of defence is to try to identify contingencies at a less complete level of detail (Approach 2), and to develop effective generic responses for these more abstractly defined situations. Lastly, for complex systems it is often impossible to know whether all possible contingencies have been predicted at either a detailed or more abstract level. In this case, a human-centred design that makes it possible for a person or team of people to effectively detect and respond to some novel scenario becomes the final layer of defence to try to ensure resilience (Approach 3).

Below, we provide a range of examples that help to further define and illustrate these different approaches to the design of resilient systems, and to demonstrate how they can be integrated in a layered fashion.

*Example 1: Resilience due to Contingency Planning in Commercial Aviation*

The narrowest and most "mechanical" use of the term resilience applies when the designer anticipates a specific kind of problem that could arise, and designs a specific solution to detect and respond effectively to this kind of problem (Smith, Stone, and Spencer, 2006; Smith, Bennett, and Stone, 2006). As one illustration, when airline dispatchers develop a flight plan, the flight planning process or "system" requires them to not only fuel the aircraft for the planned route to the desired destination but, if there is any uncertainty about the weather at that destination, to also identify and fuel for a specific alternate airport. This alternate airport is selected based on consideration of the uncertainty in the weather, and on the suitability of that airport based on safety and business concerns (Smith et al., 1997). Likewise, if isolated pop-up storms are predicted while a flight is en route, the dispatcher may add extra fuel to accommodate the added distance necessary to deviate around the predicted storm cells.

In another illustration from flight planning, an hour before departure the dispatcher may prepare for possible convective weather affecting the departure route for a flight (Smith et al., 2007). Because of the uncertainty, the dispatcher may prepare for an "envelope" indicating the range of likely reroutes in terms of their impacts on air time and fuel consumption (Smith et al., 2001). Based on how the weather has developed, a traffic manager then decides which route to assign to that flight as it prepares for takeoff (see Figure 8.1).

Each of these two illustrations demonstrates a "system" that requires contingency planning that involves:

- predicting the range of specific possible scenarios that could develop;
- providing the plans and resources necessary to deal with these scenarios;
- ensuring that the real-time decision-maker is in the loop and is capable of determining which of the alternative scenarios

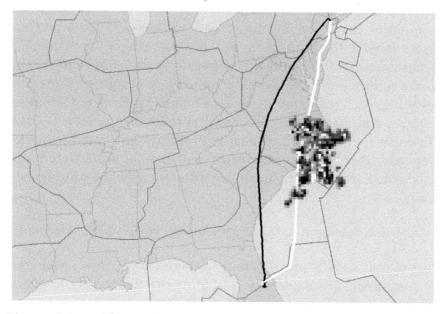

Figure 8.1    Alternative routes for departing in the face of uncertain weather

predicted by the planner has developed, responding based on the contingency plan.

Thus, in this example, the designer anticipated a scenario (flight planning in a situation where uncertain weather could impact the availability of an airport or departure airspace) and designed a "system" by defining roles and responsibilities for dispatchers and pilots and by providing decision support tools to enable this team to develop and apply specific contingency plans for each flight as needed (Smith et al., 2003).

Note also, however, that such a distributed design, with one person doing the planning ahead of time (e.g., the dispatcher) and another monitoring and initiating the use of a contingency plan (e.g., the pilot), potentially also provides a form of "layered resilience." Because the pilot is made aware of the potential cause of a problem (uncertainty about where the weather is going to develop), understands the basis and limitations of the associated contingency plan, and has access to relevant "local" real-time data indicating the nature of the situation as it arises, the pilot could also determine that the contingency plan is inappropriate for the actual situation and develop some other solution (in collaboration with the dispatcher). Similarly, because the dispatcher continues to monitor the flight and has access to data and tools that are unavailable to the pilot, there may be cases where the dispatcher determines that the planned contingencies are inappropriate and works with the pilot to develop some other response. Generally speaking, such resilience in the face of an unanticipated situation requires a human-centred design that relies on a person or team of people to recognize and deal with the unanticipated scenario. Note also that it frequently leads to plans that are different not only from the original plan, but are different from the contingency plans as well.

*Example 2: Resilience due to Contingency Planning in Medical Procedures*

Post-operative PRN (*pro re nata* or "as necessary") orders prepared by a surgeon are another example of distributed contingency planning to deal with anticipated problems, in this case during patient recovery from anaesthesia and surgery. If a foreseen problem arises, a nurse is authorized to utilize the instructions in the PRN order without further advice from the physician. PRN orders may cover medications, nursing procedures, adjustments of equipment attached to the patient, immediate notification of the surgeon or an anaesthesiologist in the event certain changes are observed either by a nurse or by monitoring equipment, and so on. As with the flight planning examples discussed above, such contingency planning makes the system more robust in dealing with predicted situations that could arise.

In addition, in this setting, one or more physicians is linked to nurses by a communication system, so that if a nurse decides something unanticipated is happening, he or she can communicate with the appropriate physicians to develop an alternative solution. As with the flight planning, this human-centred design provides a layered form of resilience, making responses possible in a timely and cost-effective manner when either anticipated or novel situations arise. If a problem has been foreseen, the nurse can deal with it immediately without the need to include other staff in the response. If not, the nurse can communicate with the physicians to identify an appropriate response.

*Example 3: Resilience in Transfusion Medicine*

Another example of resilience arises when the system designer predicts scenarios where the people operating the system could make slips or mistakes. In some cases, these disruptions are handled using software to provide the necessary safety net. In other cases, they are dealt with through the design of a distributed human system, where people (often distributed across different locations and possibly assessing the situation at different points in time) have overlapping responsibilities and can therefore detect and correct the slips and mistakes made by other operators within the system.

*Overlapping human responsibilities*   As an example, studies of blood bankers have identified a number of errors that arise in the identification of errors in determining the antibodies present in a patient's blood (Smith et al., 2008). This includes slips, as well as mistakes due to incorrect knowledge and a variety of cognitive biases (biased assimilation, ignoring base rates, perceptual and memory distortions, etc.) that lead to hypothesis fixation (Fraser et al., 1992; Smith et al., 1986).

In some blood banks, this is handled by asking a supervisor to review the blood typings generated by the medical technologists in that lab. The supervisor in essence plays the role of an expert "problem detector," quickly reviewing each case for evidence of some suspicious condition, such as a failure to explain all of the data available, or an answer that is exceedingly unlikely in terms of prior probabilities.

*Using software to detect errors* Technology can also be used to provide a safety net in transfusion medicine. As an illustration, the Antibody Identification Assistant (AIDA) provides an interactive critiquing system that incorporates an expert model of the problem-solving necessary to identify the antibodies in a patient's blood serum (Guerlain et al., 1999; Smith et al., 2008; Smith and Rudmann, 2005. AIDA uses this model to monitor human performance and to provide protection against the slips and mistakes that laboratory technologists make during the process of identifying the antibodies present in a patient's blood (see Figure 8.2).

As an illustration, a technologist might make a slip when applying a heuristic such as "rule out the antibodies against the

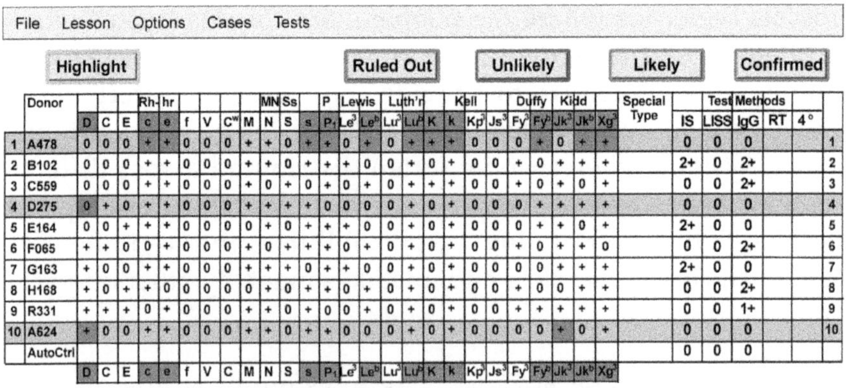


| | Donor | | | | Rh-hr | | | | | | MNSs | | | P | Lewis | | Luth'n | | |
|---|---|---|---|---|---|---|---|---|---|---|---|---|---|---|---|---|---|---|---|
| | | D | C | E | c | e | f | V | C$^w$ | M | N | S | s | P$_1$ | Le$^a$ | Le$^b$ | Lu$^a$ | Lu$^b$ | K |
| 1 | A478 | 0 | 0 | 0 | + | + | 0 | 0 | 0 | + | + | 0 | + | + | 0 | + | 0 | + | + |
| 2 | B102 | 0 | 0 | 0 | + | + | 0 | 0 | 0 | + | + | 0 | + | + | + | 0 | 0 | + | 0 |
| 3 | C559 | 0 | 0 | 0 | + | + | 0 | 0 | 0 | + | 0 | + | 0 | + | 0 | + | 0 | + | + |
| 4 | D275 | 0 | + | 0 | + | + | 0 | 0 | 0 | + | + | + | + | 0 | 0 | 0 | 0 | + | 0 |
| 5 | E164 | 0 | 0 | + | + | + | 0 | 0 | 0 | 0 | + | 0 | + | + | + | 0 | 0 | + | 0 |
| 6 | F065 | + | + | 0 | 0 | + | 0 | 0 | 0 | + | 0 | + | + | + | 0 | + | 0 | + | 0 |
| 7 | G163 | + | 0 | 0 | + | + | 0 | 0 | 0 | + | + | + | 0 | + | + | 0 | 0 | + | 0 |
| 8 | H168 | + | 0 | + | + | 0 | 0 | 0 | 0 | 0 | + | 0 | + | + | 0 | + | 0 | + | 0 |
| 9 | R331 | + | + | + | 0 | + | 0 | 0 | 0 | + | + | 0 | + | 0 | 0 | + | 0 | + | 0 |
| 10 | A624 | + | 0 | 0 | + | + | 0 | 0 | 0 | + | + | 0 | + | + | 0 | 0 | 0 | + | 0 |
| | AutoCtrl | | | | | | | | | | | | | | | | | | |
| | | D | C | E | c | e | f | V | C | M | N | S | s | P$_1$ | Le$^a$ | Le$^b$ | Lu$^a$ | Lu$^b$ | K |

| | Donor | Kell | | | Duffy | | Kidd | | | Special Type | Test Methods | | | | | |
|---|---|---|---|---|---|---|---|---|---|---|---|---|---|---|---|---|
| | | k | Kp$^a$ | Js$^a$ | Fy$^a$ | Fy$^b$ | Jk$^a$ | Jk$^b$ | Xg$^a$ | | IS | LISS | IgG | RT | 4° | |
| 1 | A478 | + | 0 | 0 | 0 | + | 0 | + | + | | 0 | 0 | 0 | | | 1 |
| 2 | B102 | + | 0 | 0 | + | 0 | + | + | + | | 2+ | 0 | 2+ | | | 2 |
| 3 | C559 | + | 0 | 0 | + | 0 | + | 0 | + | | 0 | 0 | 2+ | | | 3 |
| 4 | D275 | + | 0 | 0 | 0 | + | + | + | + | | 0 | 0 | 0 | | | 4 |
| 5 | E164 | + | 0 | 0 | 0 | + | + | 0 | + | | 2+ | 0 | 0 | | | 5 |
| 6 | F065 | + | 0 | 0 | + | 0 | + | + | 0 | | 0 | 0 | 2+ | | | 6 |
| 7 | G163 | + | 0 | 0 | 0 | 0 | + | + | + | | 2+ | 0 | 0 | | | 7 |
| 8 | H168 | + | 0 | 0 | + | 0 | 0 | + | + | | 0 | 0 | 2+ | | | 8 |
| 9 | R331 | + | 0 | 0 | + | + | + | + | + | | 0 | 0 | 1+ | | | 9 |
| 10 | A624 | + | 0 | 0 | 0 | 0 | + | 0 | + | | 0 | 0 | 0 | | | 10 |
| | AutoCtrl | | | | | | | | | | 0 | 0 | 0 | | | |
| | | k | Kp$^a$ | Js$^a$ | Fy$^a$ | Fy$^b$ | Jk$^a$ | Jk$^b$ | Xg$^a$ | | | | | | | |

**Cell #4 does not possess the D antigen. Therefore, even if anti-D was present, it could not possibly react with this cell. Consequently, this is not a good cell for ruling out anti-D.**

**Figure 8.2 Using a critiquing system to provide greater resilience in the face of a slip**

D antigen if there are no reactions (0s in the data fields) shown in the rows for IS, LISS and IgG and if Donor Cell 4 contains the D antigen (as indicated by a + in the corresponding data field)." To assure a safety net against such slips, AIDA provides a critiquing system underneath the information display, alerting the technologist if a potential slip has been detected.

Note that, once again, this is fundamentally a distributed work strategy (Obradovich and Smith, 2008). In this case, however, it is the designer of the software who is working cooperatively with the blood banker. This cooperation is occurring through the medium of the critiquing system or software.

*Example 4: Resilience in Cockpit Design*

Older instrument approach and landing cross-pointer indicator systems, or Instrument Landing Systems (ILSs) used high-frequency radio signals to move a mechanical pointer to a certain position on an instrument display in the cockpit. Two such pointers, oriented horizontally and vertically on the instrument, provided information to a pilot concerning the position of his/her airplane with respect to two radio transmitters, one providing a geographic course, and the other a "glide path" or angular course toward the ground. If one of the transmitted signals failed, the pointer associated with it became centred on the instrument, indicating a null signal. The problem with this was that a centred needle could ambiguously indicate either that the aircraft was centred on the desired course, or that the radio signal was absent.

In later display designs, "OFF" flags were added to the meter to resolve these ambiguities, but they were not always salient enough when the pilot was busy conducting an instrument approach, especially at night or in turbulence, making this attempt to increase resilience inadequate in many cases. More modern needles or symbols disappear when determined to be unreliable. This solves the ambiguity problem, and alerts the pilot that he or she must rely on other information to deal with the situation. As another example of this design approach, engine information displayed on the Airbus Electronic Centralized

Aircraft Monitoring system is replaced by an amber "X" if there is a problem with the sensor.

As with Example 1, these illustrations again demonstrate that it is important for the system designer to anticipate potential hazardous situations and design for robust or resilient responses when they occur. However, the system designer must also make sure that, when such a predicted scenario arises, the human operator will be effectively alerted so that he or she can develop an effective response (Bainbridge, 1983).

*Example 5: A Second Example of Resilience in Aircraft Design*

Many modern aircraft rely on signals from Global Positioning System (GPS) satellites for navigation. The signals from these satellites provide precise position information in three dimensions and an atomic clock that, as part of the system, provides very accurate time indications. The signals are accurate but are transmitted at very low power outputs. Early GPS variants were prone to certain types of atmospheric or deliberately induced "jamming," although the satellites now being launched incorporate improved antennas to increase signal power as well as software defences against interference.

Every eleven years Earth approaches a solar flare maximum. These solar emanations cause severe disruption of radio-frequency signals over large parts of the earth lasting for hours or days. The early-generation GPS signals could be degraded or disrupted, posing a serious navigation threat to aircraft travelling where surface-based navigation systems were not available.

Because of such potential problems, virtually all new transport aircraft have multi-mode receivers that mix various types and sources of data to provide a "best" navigation solution to the flight management system. The mix includes, as appropriate, distance measuring equipment fixes, GPS fixes and others, including inertial information. This process is transparent to the flight crew, except that the "quality" of the navigation solution is monitored; if it degrades below certain thresholds, appropriate aural and visual cautions or warnings are triggered. So, at any given time, the total navigation solution is based on combined, weighted inputs from various independent sensor platforms.

This is another variation on how redundancy can be used to provide layered resilience trying to minimize disruptions to operations as long as this is possible but providing a backup safety net (alerting the flight crew so they can decide how to deal with the situation) if the less obtrusive technological solutions are judged to be inadequate (Orlady and Orlady, 1999).

*Example 6: Layered Resilience: A Cautionary Example*

In 1994, a Tarom (Romanian) Airlines A310-300 was transporting 182 people from Bucharest to Paris Orly Airport. The flight was on final approach when it went into a steep, nose-high attitude and then rolled into a dive before the pilots regained control at 800 feet. The aircraft landed safely and no one was seriously injured.

The available data suggest that the autopilot caused the unexpected pitch-up by suddenly switching to the "level change" mode because the flap speed limit was (inadvertently) exceeded by 2 knots during the approach. However, because the automation assumed the flight should be in "level change" mode, when the pilot subsequently attempted to recover from the pitch-up, the electric trim automatically countered the pilot's actions, resulting in the dive (Billings, 1996).

This example could be construed as an example of layered resilience, with the technology responding based on one scenario that had been predicted by the designer, and with the pilot also empowered to respond, but based on a different perception of the situation. It serves as a caution regarding the need to determine which "agent" (human or technological) should have authority when there is a conflict between the two agents in a system based on layered resilience.

*Example 7: Automation and Layered Resilience in Medical Instrumentation*

Monitoring of patient vital signs (e.g., arterial blood pressure, heart rate and rhythm, respiratory rate and depth, arterial oxygen and carbon dioxide tension or concentration, etc.) has increased in importance as more seriously ill patients are attended by smaller numbers of skilled nursing and medical personnel.

Most vital signs monitors can be set, within limits, to provide (usually auditory and/or visual) alerts to medical staff when a given variable wanders outside the set limits. In some cases, these monitors can also be connected to automatic intravenous medication delivery devices programmed to provide rapid support to a patient in a contingency (e.g., extreme blood pressure, a serious cardiac arrhythmia).

Caution is necessary when such a complex of machines is able to operate unattended. Vital signs can become abnormal for any of several reasons, only some of which may be assisted by the medications provided through the automatic delivery systems. As a result, if the designer hasn't very carefully identified and dealt with the full range of scenarios that could arise, the automatic monitor could fail to differentiate among the possible causes of the phenomenon that set off an alarm, and could therefore respond inappropriately.

When used to provide layered resilience in the operating room to assist anaesthesiologists in monitoring a patient, however, such an automatic response is less of a problem, because the physician can also observe the patient directly and can look for other visual or tactile cues to a change in patient status and intervene if the automatic response is inappropriate.

*Example 8: An Example of Resilience in Military Aircraft Design*

Military aircraft (specifically, high-performance fighters such as the F-16) can exceed human tolerance limits for g accelerations, leading to transient visual blackout or even unconsciousness. Historically, crashes into terrain have been frequent enough to require the addition of autonomous control modules that sense the imminence of a crash and institute a terrain avoidance manoeuvre automatically in order to give the pilot time to recover his or her senses safely. These are "last-ditch" safety devices that cannot be overridden by the pilot.

This is an example where a specific cause (pilot blackouts due to g accelerations) has been anticipated by the designer, and a specific technological solution has been implemented.

*Example 9: A Second Example of Resilience in Military Aircraft Design*

Norris (2007) reports:

> The U.S. Air Force has successfully completed the first separation tests of the GBU-39 small diameter bomb [SDB] from ... a Lockheed-Martin F-22 as part of the fleet-wide Increment 3.1 upgrade program. ... The team also revealed that during SDB tests a maneuver inadvertently caused a momentary double-engine stall. The incident ... was triggered when the pilot, flying inverted, was checking pre-set negative trim adjusted to -1g to enable smooth entry into an inverted roll. ... When a second run was made to check a different setting, also inverted, the F-22 briefly encountered a zero g condition which caused a "bubble void" in the fuel intake.
>
> Although the momentary stall was apparently undetectable by the pilot, the power interruption cut telemetry, hot microphone and radio links to the ground test room. "The screens just froze," says the team, adding that the last thing the ground operators saw was the aircraft going down, very fast and very low. The F-22 self-recovery system responded correctly, members of the team pointed out.
>
> To avoid any recurrences, the Combined Test Force says all trim checks will be made while flying upright, and a redundant fuel pressure monitoring system has been set up in the control room.

*(p. 35)*

Based on this report, it appears that this is a nice illustration of a case where the "cause" (a bubble void in the fuel tank) was not anticipated by the designer, but the possibility of an undesirable "effect" or state (the plane "going down, very fast and very low") was predicted. Detection of this undesirable state was used as a trigger for automation, providing a resilient response even though the cause had not been anticipated during the design.

## Conclusion

The preceding examples serve to highlight a number of important design concepts relevant to increasing resilience. First virtually all of them incorporate some approach to *layering*. Examples 1 and 2, for instance, illustrate systems where operational staff (airline dispatchers or physicians) attempt to predict possible scenarios that could arise, and prepare for the implementation of explicit *contingency plans* to deal with these situations should they arise (Hollnagel, 1993). As a second layer of defence, people are positioned to detect anomalies that are not covered by these contingency plans, and to develop and implement solutions on the spur of the moment. As illustrated by Examples 4 and 5, however, to be effective this layer of defence requires the design of *informative, salient alerts* that focus the person's attention on the problem.

In the aviation example, this detection task is strengthened by designing *redundancies* into the system. Dispatchers and pilots (as well as FAA traffic managers and controllers) are all monitoring in real time in order to detect situations that require some adaptation of current plans. And because expertise and access to data is distributed across these different people (across dispatchers and pilots in the aviation example and across nurses and physicians in the medical example), this second layer is further enhanced by providing pre-defined *communication channels to support collaboration.*

Example 3 extends this use of redundancy to increase resilience through the introduction of a critiquing system as the collaborating agent in blood banks. Strictly speaking this is still a form of human-human collaboration, as it is the software designers who are collaborating with the blood bankers through the medium of the computer.

Example 3 further emphasizes that, to be resilient, systems must not only deal [with] anomalous changes in the environment, but must also provide *safeguards against slips and mistakes* made by people (Fraser, 1992). These slips and mistakes could be made by the real-time "operators" in the system (e.g., the blood bankers completing the antibody identifications), with the software (or more correctly, the software designers) providing a safeguard in the form of a critiquing system. They also could arise when the software designers make a slip or mistake and fail to anticipate some scenario, leading to brittle performance by the software (Leveson et al., 2006; Reason, 1991).

In this latter case, by assigning a role that helps ensure that a person (e.g., the blood banker) is engaged in the task in real-time, and by providing sufficiently perspicuous representations or displays so that this person can detect the onset of novel or unanticipated scenarios that exceed the software's competence, this person provides a *safeguard against the brittleness of the software.*

This synergistic relationship can be made even more effective by providing the software with *metaknowledge*, or knowledge about the limits of its knowledge, thus allowing it to detect and alert the person when a scenario arises that is outside of its range of expertise (Smith et al., 2008).

Example 6 provides a good example of brittleness in an advanced decision support system. That example highlights the importance of determining *which agent (human or machine) should have authority*, in order to avoid conflicts between the human operator and the software.

Examples 7, 8, and 9 contrast nicely with Example 6 in which the human pilot ultimately averted a crash, in the sense that there are times when an immediate response is necessary to deal with an imminent danger (a plane crashing or a patient dying), and where it may be desirable to *empower the software to make an instant response in order to stabilize the situation.*

Examples 8 and 9 also serve to bring us back to the themes introduced at the beginning of this chapter, namely that there are different complementary approaches that can be considered during the design process in order to increase resilience. In Example 8, a *causal chain was anticipated* à la Approach 1 (g forces render the pilot unconscious, who is therefore unable to pull the aircraft out of a dive), and dealt with through the software design (giving the software the authority to take over control and avoid the terrain). In Example 9, the cause of the problem (a bubble void in the fuel intake due to a zero g condition when the aircraft was flying inverted) was never anticipated, but because the software was programmed to *intervene based on the state alone* (the aircraft "going down, very fast and very low"), a crash was averted à la Approach 2. And finally, many of the examples illustrated hybrid designs (Approach 4) in which more than one approach was applied, so even when one approach was inadequate, another provided the necessary resilience.

# PART IV
# Applications and Implications

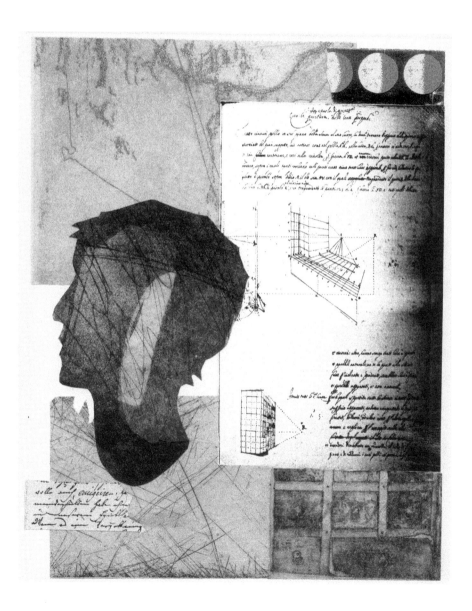

# Chapter 9

# Notes from Underground: Latent Resilience in Healthcare

Shawna J. Perry
Robert L. Wears

## Introduction

The manifestations of resilience can be varied and are often difficult for organizations to articulate or officially recognize. For example, healthcare organizations commonly respond to risk and hazard by prescribed responses based on ideas about how work *should* occur (i.e., how it is imagined). This leads to an over-reliance on policy and procedures for determining what are safe and usual operations, and indirectly to a devaluing of factors contributing to the emergence of resilient tactics and performances (Perry et al., 2007b). This often leads to informal systems for both performing work and for informing workers about hazards, which function alongside but relatively independently from the formal, "official" systems (Cross and Prusak, 2002; Freilich, 1991).

This chapter will discuss how an informal communications network served as a latent manifestation of resilience in a large healthcare organization, and how such networks can contribute to and degrade safety.

## Background

In order to provide context, we briefly describe a specific equipment problem that highlighted a little-known and hard-to-see network of resilient interactions.

*The Setting*

Recent advances in the management of diabetes require more frequent measurement of blood glucose levels to maintain it within narrow limits. Hand-held devices, called "glucometers," are used for bedside measurement of blood glucose (see Figure 9.1). The glucometer measures the blood glucose from a single drop of blood placed on a testing strip inserted into the device which displays the glucose level on its LCD screen in less than one minute. Because blood glucose results from the main laboratory typically require a physician order and it can be hours being reported, bedside glucometer testing has become extremely popular.

The setting for these events was a 500-bed, inner-city, academic medical centre with 220,000 annual visits and over 3,200 employees. Federal regulation for bedside or "point of care" testing (e.g., pregnancy tests, urinalysis) has resulted in the organization taking steps toward more direct management of these activities. The organization responded by creating a new division with dedicated staff to monitor point-of-care testing, procedural guidelines for users, and an information technology (IT) based system for monitoring bedside glucose testing with handheld devices. This system required barcode scanning of the operator's name badge for each operation, allowing it to track individual users. Thus, access and use of the devices is centrally controlled by laboratory services and the office of point-of-care testing. Over 40,000 bedside glucose tests are performed at this facility each month.

The hospital also uses a commercially available IT-based hazard reporting system to accumulate and track information on near misses, incidents, hazards, and accidents. The system has been in use since 2004 and has

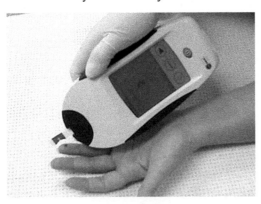

**Figure 9.1    A typical hand-held glucometer**

been regularly upgraded. Information can be entered into the system by any staff member, although a login ID and password are required. Entries are reviewed daily by the hospital's Risk Management and Patient Safety Office. The system has averaged over 2,500 reports per year.

The hospital patient safety officer is a former nursing director with 25 years of experience in the organization. She has provided nursing care and administrative leadership on several different clinical units, including the general wards, intensive care unit and the emergency department. Because of this history, she has developed an extensive network of personal and informal contacts amongst clinical and administrative staff members and is generally regarded as a reliable, trustworthy and discreet resource.

*The Problem*

Over a two-week time span including a major holiday, three incidents were identified in which the blood glucose measurements reported by handheld glucometers did not comport with the patient's condition (Perry et al., 2007a). The nurses involved reacted by taking several measurements on different glucometers (borrowed from other units) in an attempt to reconcile the results displayed and the patients' conditions. Per hospital protocol for suspected aberrant readings, simultaneous blood glucose levels were also sent to the hospital's central laboratory for confirmation. Results showed the handheld measures to be falsely high on several occasions, and falsely low on others. These aberrant measures came close to causing patient harm by leading to inappropriate treatment or delaying appropriate treatment, although there is no evidence that any patient suffered permanent harm, as in each case the nurse involved opted to act based on the clinical findings alone.

An investigation was initiated by the patient safety officer who heard about the glucometer problems described above during informal conversations with nursing supervisors, managers and emergency department (ED) personnel. A total of seven incidents of aberrant glucometer results (including the three discussed above) were verbally reported to the patient safety officer.

The nurses involved in all three cases were experienced with the devices and stated they had received no indication of user error or malfunction. All were concerned by the marked disparity between the bedside glucometer readings and the confirmatory values. They were, however, not surprised by the occurrences themselves: "We know they can happen, that's why we check on other machines when we think ours is wrong" (Perry et al., 2007a).

The erroneous handheld glucometer results were temporally related to an unscheduled change in lot numbers for the reagent testing strips used for bedside testing; however, a number of additional variables were identified as possible contributors (e.g. blood sample contaminated with skin debris; incomplete sample size). The exact reason(s) for the erroneous measurements has yet to be completely elucidated.

Only two of the 2,250 reports made to the organization's electronic reporting system during the six months before and after the near miss incidents related to the glucometer, and they were posted after the problem had been passed to the patient safety officer via the informal network. The laboratory and point-of-care office were unaware of any of the incidents until contacted personally by the patient safety officer.

*Analysis*

There is no mistaking the contribution of human factors and device usability to the confluence of factors that resulted in this cluster of cases (Perry et al., 2007a). This chapter, however, will focus on: (1) the paradoxical role played by the organization in undermining its own resilience as it attempted to enhance safety; and (2) the emergence of latent mechanisms for preventing its disappearance.

## The Organization as a Source of Failure

The creation of an office and dedicated staff to manage bedside testing was an unmistakable demonstration of the organization's commitment to meet growing regulatory demands for demonstrable accountability. However, the organizational division of labour (in assigning this function to the laboratory) contributed in some ways to the subsequent problems. The laboratory and clinician perspectives of the work of healthcare are quite different. Laboratory work tends to be more task-oriented (e.g., blood analyzed and results reported, regulatory standards met, etc.), while clinicians tend to alternate attention between tasks and goals, in particular favouring goals over tasks in cases of conflict (i.e., the goal of mitigating illness or injury may sometimes require a deviation or violation in the way in which tasks are performed). Manifestation of this conflict of perspectives can be found in a number of procedures for tightening control over bedside testing, which included:

- the requirement to scan a user's badge in order to activate the glucometer and enforcing this requirement in the software;
- "locking out" users from the glucometer for deviations from procedure (e.g., scanning the barcode identifying the patient before calibrating the machine);
- reactivating access to the device following each lockout only after attending remediation classes.

This privileging of tasks over goals resulted in increased system brittleness around glucometer use and reduced resilience, especially in situations of "off normal" operations. The implementation of an IT system for tracking compliance resulted in a shift in location of the point where conflicts between goals and tasks were resolved. Previously resolved locally at the point of care by a caregiver in context, conflict resolution is instead performed by designers or managers in advance and out of context, enforced by software and organizational procedures. This change in focus favoured one set of perspectives and values over another, in ways that were not clearly understood or negotiated by the participants at that time.

Similar unanticipated shifts between competing demands have been noted in other areas of clinical care in many healthcare institutions. For instance, denial of computer access to clinicians until delinquent medical records are completed, or in emergency situations, requiring the completion of a patient's registration in the hospital information system before the software will allow input of orders for laboratory tests or access to x-ray results. The global "locking out" of clinicians and staff for procedural non-compliance without regard to contextual urgency of the violation prioritizes the organization's goal of compliance with external regulatory and reimbursement rules over the clinicians' immediate information needs for patient care (Griffey et al., forthcoming).

Healthcare organizations typically have a narrow view of safety and reliability, functioning in what has been called an "anticipatory model" (Flach, 2003; Schulman, 1993a). This organization's actions, favouring compliance over context, are illustrative of Schulman's observations of managers attempting to "lock in" organizational performance through elaborate rules and procedures, formal authority assignments, and clearly differentiated job responsibilities. The hospital's creation of the office for bedside testing, with the assignment of personnel for management, education and remediation of glucometer usage, reinforces the false assumption that "perfect work performance" will create safety. This is likely due to unrealistic expectations of what safety, reliability, and resilience are, and a philosophy of improvement through increased standardization and centralized control that is pervasive in healthcare (Timmermans and Berg, 2003). This contrasts sharply with the view that safety, reliability, and resilience cannot be achieved through attempts at controlling procedural invariance (e.g., the rigid enforcement by technology of policies and procedures) but rather through the situated management of fluctuations in organizational relationships and work practices. This strategy enhances reliability while preserving the protective functions of organizational slack.

There were other factors that undermined the organization's overall glucose management system but these received much less attention under the narrowed anticipatory model of safety and reliability. The bedside testing technology in use seemed to perform reasonably well under ideal circumstances, but was found to be highly intolerant of seemingly minor deviations in procedures. It is almost impossible to control or track the effect of these minor deviations on clinical care because of throughput or production pressures and workload demands. Additionally, advances in management of diabetes have increased pressure for tighter control of blood glucose levels, resulting in more frequent testing. The result has been increased opportunities for failure from minor deviations in what is generally considered by hospital staff to be a benign procedure.

## The Human Factor

Patients and staff were important sources of resilience in these seven glucometer incidents (Anders et al., 2006). The staff recognized the mismatch between data received from the glucometer and the clinical picture before them and sought confirmatory data by testing other devices. In one of the cases, the patient was an added source of resilience as she insisted that the device readings could not be correct, and sought to prove it by having her spouse perform readings on her personal glucometer.

The resilient behavior demonstrated by the staff and patients in these cases support other models of performance, such as the "resilience model" (Hollnagel, Woods, and Leveson, 2006; Schulman, 1993a), where being responsive to rather than weeding out the unexpected is a more suitable strategy in uncertain environments. The flexibility of the humans involved in these cases and their adaptive capacity when the limits of the glucometer protocol failed, mitigated the risk of failure or harm. This is particularly significant because none of the actors knew what the cause(s) of the false glucometer results were, and as such, had very weak data from which to act in a resilient manner. Managing to achieve invariance in performance in such healthcare settings threatens the extemporaneous emergence of resilience, and thereby the dynamic creation of safety (Leveson, 2003).

## Notes from Underground

An intriguing aspect of this cluster of seven near misses is the relative paucity of formal reports related to them in the hospital's electronic reporting system, the designated and expected channel for reporting safety issues (Wears, Perry, McDonald, and Eisenberg, 2008). How knowledge of this problem was represented and distributed in the formal and informal communications networks provides a demonstration of resilience within the organization that is off its radar and obscured from detection.

*The Formal Reporting Structure*

The organization's formal reporting structure is represented in Figure 9.2. Clinical operations that frequently share safety information via the informal system are shown with a bolded border. Non-clinical functions (e.g., pharmacy, biomedical engineering, and campus safety/parking) that frequently report to the electronic reporting system are shown by dotted boxes at lower right. The safety officer's position is not shown on the official organizational chart, but has been added to the figure (indicated by the star) for the purposes of this discussion. Note that the organization structure is largely hierarchical, and the safety officer is not only absent from the official functionality chart but is also outside of the mainstream of organizational or clinical work.

*The Informal Reporting Structure*

The patient safety officer learned of the initial three cases through an informal reporting system of apparently casual conversations

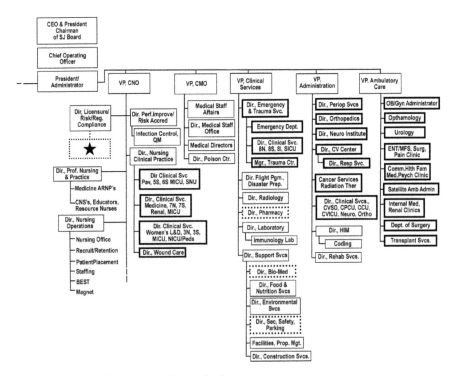

**Figure 9.2    Organizational chart**

with nursing supervisors, managers and ED personnel. Interviews with the patient safety officer found the electronic reporting system being bypassed, with information on any incidents of concern being directly communicated to her through an informal social network. The majority of these interactions were verbal, conveyed in conversation or voicemail and, on occasion, via email. The shaded ovals in Figure 9.2 represent the clinical areas that are key contributors to the informal network (e.g., in-patient medical wards, intensive care units, surgical services, etc.). Table 9.1 lists some of the positions and functions frequently involved in reporting via the informal system, and the approximate frequencies of reporting.

The computerized reporting system in the organization is intended to direct the flow of safety-related information via official channels toward the safety office. The unshaded ovals in Figure 9.2 indicate significant non-clinical contributors to the hospital safety reporting system from pharmacy, biomedical engineering, and campus safety/parking, each of which provide occasional informal verbal reporting directly to the safety officer. Their contribution to the informal network strongly suggests a recognized effectiveness for filtering safety concerns up the organizational tree for greater attention and possible resolution.

**Table 9.1**     Contacts and contact frequency for informal reporting system

| Reporting contact | Frequency of contact |
| --- | --- |
| Nursing supervisor | Daily, 3 shifts/day |
| Rapid response team nurses | Daily, multiple time |
| Pharmacy | Daily, multiple times |
| Facility safety manager | Daily |
| Risk managers | Daily |
| Aides, techs, unit secretaries | 3–4 x per week |
| Staff registered nurses | 3–4 x per week (regular rounds) |
| Vice President facilities | Occasional |
| Biomedical engineering | Weekly |
| Nurse educators, case managers | Weekly |

The extensive, informal contacts within this network, many initiated as a consequence of the safety officer making herself available without respect to the formal organizational structure, appears to facilitate the flow of safety information in the organization. Cross et al. described this type of social network as a "customized response network," best suited for solving ambiguous problems requiring rapid problem framing and coordination of the relevant expertise to find resolution (Cross et al., 2005). A customized response social network is collaborative within and across organizational lines with expectations of reciprocity by all involved. It may make it more likely for unpopular or potentially threatening information to be recognized.

A contrasting social network described by Cross is the "routine response network," which is suitable for solving familiar and routine problems with known responses because the work is essentially standardized, such as in a bank or an insurance company. The expectation of healthcare organizations and regulatory groups is that information regarding hazards will flow thorough routine pathways, following a type of chain of command. Despite the presence of a sanctioned reporting structure, the hospital in the example above developed an underground social network that obviated the formal expected methods of reporting, resulting in a latent "grass roots" system for safety-related information. It is an ongoing type of resilience that reports unpredictable hazards and "real work" risks that must be attended to as quickly as possible if safety is to be maintained.

## Strengths and weaknesses of the informal network

The existence of mismatches between reality and formality impedes organizational learning and may lead to policies or procedures that are ineffective or even harmful (Fujita, 2006; Perry et al., 2007a; Tucker and Edmondson, 2003). The existence of an underground network as a latent monitor for safety in parallel with (and in spite of) the formal computer-based reporting system highlights some of the problems of safety reporting in organizations. In this example, information on a hazard appeared more quickly, in greater volume and with greater detail than it did in the formal safety reporting system. There are several possible reasons for this.

First, the informal network is truly voluntary, is more easily accessed (e.g., opportunities to report arise during the routine course of work), and appears safer to workers, who are able to speak freely about safety concerns to someone trusted. In contrast, the formal reporting system requires workers to obtain logins and passwords, which link each report directly to reporters. Using the informal network protects reporters from the "blame and re-train" aspects of formal reporting by buffering reporters from problems.

Second, the informal network affords an opportunity for interaction, while information flow in the computerized system is all one-way. Thus, workers can raise issues without having to be sure about what happened, and they can get immediate positive feedback for reporting, providing reinforcement for their behavior.

Third, the interactive nature of the informal system gives workers the ability to "tell a story" about an event, rather than having to fit it into the formatted structure pre-specified by the software. This allows rich, thick, narrative descriptions, including many of the "messy details" important for understanding hazards and how they are dealt with (Nemeth, Cook, and Woods, 2004).

Fourth, the success of the informal network likely depends on a pre-existing history of cooperation between the participants and the safety officer. Participants volunteer hazard information in anticipation of reciprocity; based on past experience, they expect that they will not be exposed, and that they may benefit in the future from information flowing back to them. Because of the lack of official organizational structure providing a clear remit for the safety officer's role and activities, it seems unlikely that such a network would function as successfully without pre-existing personal relationships.

However, there are some disadvantages to the informal reporting system. First, the lack of transparency of the network, with its "behind the scenes" stance, increases the risk of bias and selective response to the incidents relayed. There can also be encapsulation of the reports, resulting in incidents and issues not moving beyond the network to those within the organization who are empowered to respond to them. Additionally, the individual who receives the information (the patient safety officer) is not well positioned to take action nor are most of the individuals in the network. Nurses play a prominent agentive role in the maintenance of this informal network, consistent with their role within healthcare as ad hoc advocates for safety. The deconstruction of nursing culture in the United States reflects the limited recognition of their full role and hence a devaluation of their contribution to monitoring risk and creating safety (Weinberg, 2003).

## Conclusion

The informal safety reporting network discussed here is an important mechanism for sharing safety information, but ironically, can also encapsulate and potentially hide it. Changes in clinical work occur continuously, and front-line workers often recognize the impact of such changes but may have difficulty articulating this knowledge to the organization. At the same time, managers at the "back end" tend to have sanguine ideas that operations always proceed as planned, and that rules and procedures can keep the system operating in an optimal way (Fujita, 2006). This mismatch constitutes a fundamental type of latent failure in complex organizations which the underground network of safety communications has the potential to remedy, if it can be successfully tapped without destroying the very features that make it work.

# Chapter 10

# Cognitive Underpinnings of Resilience: A Case Study of Group Decision in Emergency Response[1]

David Mendonça
Yao Hu

## Introduction

Observations of group decision at the frontier of human experience promise insights into how human collectives anticipate and respond to highly non-routine events. In the case of decision-making by emergency response organizations, prior experience is expected to be relevant, despite the sometimes considerable difference between those experiences and the emergency situation at hand (Earley, 1985; Salas et al., 1992). Understanding how knowledge gleaned from these experiences is used (or misused) in restoring services disrupted by emergencies should deepen our insights into how collective creativity and joint expertise may contribute to organizational resilience.

This chapter reviews work in developing models of the cognitive processes that underlie decision-making by emergency response organizations following the onset of highly non-routine situations. The particular type of incident considered here is an industrial accident, where a breach has occurred in the engineered system (in this case, a ship). The focus of the chapter is on explaining the impact of event severity on cognitive and

1     This material is based upon work supported by the US National Science Foundation under Grant Nos. CMS-9872699 and CMS-0449582. The authors thank the US National Fire Academy for helping to make this work possible.

decision processes among a group of response personnel as they dispatch resources to mitigate the effects of an emergency event of this type.

Background is first provided on the expected impact of event severity on divergent and convergent thinking (i.e., the processes thought to characterize creative cognition). Specific research questions address the impact of event severity on the relationship between alternative solutions considered by the group, recommendations of group members regarding decisions to be made, and finally decisions themselves. Data from one group is then examined in considerable detail. Results of the study suggest how group decision processes contribute to – or detract from – resilient performance following disaster.

## Background

Emergencies present unique opportunities for testing the limits of explanatory theories of human cognition and decision. By definition, emergencies occur at – and sometimes far beyond – the boundaries of human experience, where creativity and an ability to decide effectively against time constraint and in the presence of high stakes are highly valued (Kreps, 1984; Mendonça and Wallace, 2007). As emergencies increase in scale, they are likely to be addressed by multi-disciplinary groups. Yet very little research has been done into the link between cognition and decision-making among personnel in emergency situations (Mendonça and Wallace, 2004).

Group work is common in the response to large-scale emergencies, with members likely to be drawn from disciplines such as fire, police, health and hazardous materials management (Scanlon, 1994). Members of the group (typically called an Emergency Response Organization, or ERO) are expected to work collectively to consider various courses of action, make recommendations to a group coordinator on how to manage the emergency, and communicate decisions to field commanders about personnel and material resources to be sent to the scene of the emergency. Once resources arrive at the scene, they are managed by on-scene commanders, who communicate as necessary with the ERO (Mendonça and Wallace, 2007).

## Resilience in EROs

By definition, EROs expect to be subjected to shocks both to their own system and to the systems they manage. Resilience has been defined as an ability to resist disorder (Fiksel, 2003), as well as an ability to retain control, to continue, and to rebuild (Hollnagel and Woods, 2006). In the context of ERO operations, disorder may arise when planned-for ERO staff (Kendra and Wachtendorf, 2003) or resources are unavailable; control may be challenged when lines or organizational authority are blurred or underspecified. The following factors are thought to contribute to resilience (Woods, 2006):

- *buffering capacity*: size or kind of disruption that can be absorbed/ adapted to without fundamental breakdown in system performance/structure;
- *flexibility/stiffness*: system's ability to restructure itself in response to external changes/pressure;
- *tolerance*: behaviour in proximity to some boundary;
- *margin*: performance relative to some boundary;
- *cross-scale interactions*: how context leads to (local) problem-solving; how local adaptations can influence strategic goals/ interactions.

For simplicity, the system under consideration here is the ERO itself, acting in response to a (simulated) emergency situation. The focus of this study is on decision-making by an ERO during a simulated, non-routine, time-constrained situation. Accordingly, performance here refers to the efficacy of the ERO's decisions. Behaviour refers to cognitive and decision-making processes within the group, here denoted *solution assembly*. The structure of the group refers to how the group organizes solution assembly processes, as well as to how roles are performed by group members. As will be clarified later, cross-scale interactions are not considered since action is confined to the ERO interacting with the simulation (as opposed to, for example, the ERO interacting with strategic-level decision-makers).

Creativity has long been viewed from a theoretical perspective as entailing both divergent and convergent thinking processes.

Divergent thinking involves "multiple or alternative answers from available information," while convergent thinking involves deriving "the single best (or correct) answer to a clearly defined question" (Cropley, 2006). However, the vast bulk of the literature on creativity in groups has treated the two types of processes as distinct and, perhaps, separable (Kerr and Murthy, 2004). Yet, as is clear from Weick's study and those of others (Cropley, 2006), convergent and divergent processes may operate in tandem.

Previous work has suggested that high levels of time pressure and event severity may cause EROs to choose familiar courses of action even when they do not match field conditions (Weick, 1993). As with various other groups, it is appropriate to consider the impact of external conditions on how the ERO responds to highly non-routine situations in order to gain insight into how groups think (or fail to think) creatively (Hinsz et al., 1997). Accordingly, the research questions developed in the next section examine solution assembly processes in group performance in relation to factors thought to contribute to resilience.

## Research Questions

This study examines solution assembly by a multidisciplinary group responding to two simulated, highly non-routine emergencies, each of a different level of severity. Consistent with Social Decision Scheme Theory (Stasser, 1999), the design of the study follows from the view that group decision is best investigated as the joint product of task structure, group practices, and properties of the environment. The task of the group here is to make joint decisions about which resources to allocate to an emergency scene. The practices of the group are aligned with the conventions of EROs. The environment is one characterized principally by time constraint and event severity.

The analysis is organized around five research questions.

*Question 1*: How does the effort devoted to solution assembly change over time, as reflected in the evolution from considered courses of action to decisions? As discussed below, the difficulty of the situation facing the ERO here increases (sometimes unexpectedly) with time, thereby introducing disruptions. This question is therefore relevant to *buffering capacity*. It is expected that effort will increase with time, until the point where considered courses of action (CCAs) begin to become infeasible, after which it will decrease. Increased level of severity is expected to increase level of effort, but not the trend. It is also expected that decisions will be made close to (but not after) the point at which resources begin to become infeasible.

*Question 2:* To what extent do the recommended uses of resources for given response goals deviate from convention? The situation facing the group is characterized by various non-routine events, such as unexpected unavailability of equipment and personnel. To succeed in such situations, groups must be able to demonstrate flexibility (Mendonça, 2007), here conceptualized as the extent to which they consider unconventional uses of available resources. This question relates to the *flexibility/stiffness* within the ERO. Groups are expected to be less likely to consider unconventional uses of resources (e.g., using pumper trucks for controlling access to the incident location) when event severity is higher (Kerr and Tindale, 2004; Weick, 1993).

*Question 3:* To what extent do CCAs explain variability in recommendations of individual group members? Individual group members are free to make recommendations on how best to use existing resources, and it is assumed that there is some connection between CCAs and these recommendations. It is perhaps useful to consider the set of CCAs as forming a type of boundary (in the sense of a consideration set). Because the task of the group is cooperative, it is expected that recommendations will be strongly informed by CCAs (as opposed to unexpressed preferences of individual group members). This question therefore reflects upon *tolerance.* CCAs are said to explain recommendations if resources included in CCAs tend to appear in recommendations. It is expected that frequency of mention of a resource will increase the likelihood that it is included in a recommendation. The effect is expected to be smaller as event severity increases.

*Question 4:* To what extent does the size of CCAs (as reflected in the number of resources used) mirror that of decisions actually taken? As in Question 3, the issue is the extent to which the group can work within the (soft) constraints that arise as a result of a consideration set (i.e., CCAs) being formed. Group members may offer CCAs that can be combined into larger, potentially more efficient, decisions, but this type of activity requires more work than generating CCAs. This question therefore addresses *tolerance*. It is expected that the size of considered courses of action will parallel that found in decisions actually taken, since the decision by the group is meant to be a collective one. It is expected that event severity will have no appreciable impact on this effect.

*Question 5*: How does the set of CCAs impact decision quality? Decision quality in the context of emergency response operations is a multi-faceted construct: it concerns avoidance of human and economic losses, evaluated with respect to resources of time and material available to the group. This question is most relevant to the *margin* factor, since it concerns both performance and behaviour. It is expected that, as the number of considered courses of action increases, better solutions become more likely (Kerr and Tindale, 2004). Event severity is expected to impact this relationship by reducing the number of considered courses of action, and thus the likelihood that a good solution will be found (Weick, 1993).

## Method

### Experiment Design

The group was first presented with an overview of the session, then completed consent forms and background questionnaires. They next practiced using the computer-based system on a test case until all stated that they understood their task and how to use the computer-based system to perform it. They were then randomly assigned to the experiment conditions (here, the low-severity Case 1 followed by the high-severity Case 2). After each case, they assessed their performance. To conclude the experiment, they provided feedback via questionnaire response and informal discussion on the experience of participating in the session.

### Participants

The five participants were emergency managers who were taking part in professional training at the US National Fire Academy (NFA). These participants were solicited through presentations made to emergency managers enrolled in an NFA course in incident management. Their participation was voluntary and not linked to the course itself.

One participant in the group served as group coordinator, with each one of the other participants serving as the representative of a particular emergency service: Police Department (PD), Fire Department (FD), Medical Officer (MO), and Chemical Advisor (CA). Members self-selected these roles, and all reported that they felt comfortable taking on the role for the purposes of this exercise. A summary of prior experience and qualifications of participants is given in Table 10.1.

**Table 10.1    Summary of participants' experience**

| Role | CO | FD | PD | MO | CA |
|---|---|---|---|---|---|
| Age | 46 | 40 | 43 | 50 | 49 |
| Highest level of schooling | College/ University | College/ University | College/ University | College/ University | College/ University |
| Major | Communications | Education/ Business | Fine Arts Teaching | Public Admin | Public Admin |
| Year | Graduate | Masters 1983, 1985, 1993 | Graduate school final year | Masters | Graduate |
| Specialized training in | Incident Command System, Hazardous Materials, Technical Rescue | Hazardous Materials, Confined Space Rescue | Incident Command System Leadership Tactics, Paramedic | Safety | Emergency Medical Services, Fire, Hazardous Materials |
| Present job title and function | Battalion Chief – Incident Mitigation Prevention, Emergency Management | Chief Training Officer | Assistant Chief Administration | Safety Officer | Fire Chief/ Administrator |
| Emergency response experience | fire, press relations, hazardous materials | fire, press relations | fire, press relations, hazardous materials, medical | fire, hazardous materials, medical | fire, press relations, hazardous materials, medical |
| Past career positions | Fire Captain, Fire Lieutenant, Firefighter | Firefighter, Driver/ Operator, Captain, District Chief | Firefighter, Driver, Lieutenant, Captain, Shift Commander | US Marine Corps, Firefighter, Captain, Fire Inspector, President of Fire Relief Association | Firefighter, Paramedic, Chem/Bio, Training, Captain, Shift Commander, Fire Chief |
| Years in present position | 4.5 | 3 | 6 months | 5 | 11 |

**Table 10.1   *Concluded***

| Role | CO | FD | PD | MO | CA |
|------|-----|-----|-----|-----|-----|
| Years with present organization | 27 | 15 | 1 month (previously, 23 years) | 24 | 26 |
| Years experience working with computers | 8 | 20 | 20 | 5 | 16 |
| Number of emergency response exercises in which you have participated | >15 | Too many to count | 400 plus | >30 | >100s |
| How many actual emergency responses have you been involved in | over 1,000 fires, hazardous materials, over 1,000 medical emergencies | Too many to count | 46,000 plus | >10,000 | >10,000 |

The group convened in a conference room, seated so that they could view each other face-to-face. Video cameras and microphones recorded their interactions (see Figure 10.1).

Before taking part in the experiment itself, participants took part in a training session to familiarize themselves with the computer interface (described below). They undertook and reviewed a single practice case until they all stated they understood the interface and the task.

*Cases*

The group was presented with two cases, developed by the lead authors and others (Mendonça et al., 2006) from archival reports of prior accidents, supplemented by extensive discussions with subject matter experts at the emergency services division of a major European port. Case 1 concerned a cargo ship fire with subsequent oil spill; Case 2 concerned a collision between two ships with subsequent chemical emission. A conscious effort was made in developing the cases to create situations which were plausible but non-routine. In practice, this meant providing a set of standard resources (i.e., those typically available to the services) that were

**Figure 10.1    Experiment setup**

sufficient for a minimum acceptable-level response. However, information was differentially distributed across participants. The lack of complete information for any one participant (except the coordinator) required the group to communicate in order to understand which resources were available. The cases were made non-routine by including plausible but rare constraints on the use of resources (e.g., $CO_2$ canisters that have to be escorted via police vehicles), availability of resources (e.g., some vehicles did not have drivers), and uncertainty in information provided to the group (e.g., number of passengers aboard the ship was unknown). These constraints created a more complex working environment, further forcing communication and collaboration.

Each case had two phases. Phase 1 of each case required the group only to plan the response to the current event situation. Phase 2 followed, and began with a report that the situation had escalated and that none of the orders for dispatching resources to the incident location had been executed. Moreover, certain resources which had been available in phase 1 were not available in Phase 2, while other resources had been added. The added resources were so-called alternative resources (AR): that is, resources which were somewhat unconventional for the incident, such as gravel trucks and helicopters, but which were available for use by the group. Finally, it should be noted that, in Phase 2, the group had to account for the time required for vehicles to travel to the scene (indicated on the map). Consequently, soon after the commencement of the case, resources began to become infeasible since they could not reach the incident location within the deadline. Therefore, every minute spent on plan development was a minute that was unavailable for plan execution.

Goals G1–G4 were given in Case 1 (C1), and G1–G5 were given in Case 2 (C2):

G1: control of access to incident location
G2: control of fire at incident location
G3: removal of trapped persons from danger
G4: treatment of injured persons
G5: control of chemical release.

Information on the cases was presented through a computer interface (see Figure 10.2) with a map on the left side showing the location of available resources and panels with information about the status of the emergency on the right. A service representative learned about resources controlled by his or her own discipline by clicking on sites controlled by that service (indicated by icons on the map). Information about other disciplines' resources had to be requested verbally. The Coordinator CO had access to all information. A log entry was generated each time the participants interacted with the computer-based system.

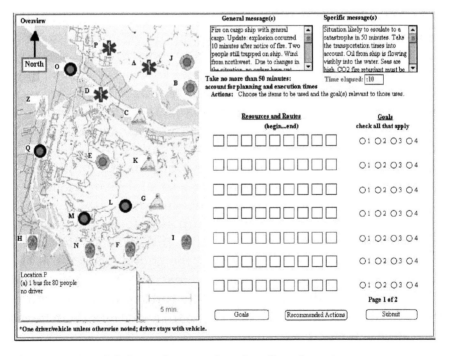

**Figure 10.2    CO interface to the simulated system**

Figure 10.2 may also be used to illustrate the nature of the time constraint in Phase 2 of each case. Note that site *I* is approximately 23 minutes (via shortest path) to site Z. Consequently, any decision involving resources at site I had to be executed (i.e., submitted by the coordinator) before 27 minutes had elapsed since the onset of Phase 2.

Roughly the same number and mix of resources were available in each case, with up to three resources (denoted *a*, *b*, or *c*) available at each site. For example, in C1, resource *Aa* was an ambulance located at site *A*. Table 10.2 lists all resources initially available at all sites.

**Table 10.2    Sites and corresponding resources for cases 1 and 2**

| Case | Sites | Resources A | B | C |
|---|---|---|---|---|
| 1 | A | 1 ambulance | 1 ambulance | |
| | B | 1 pumper truck | 1 pumper truck | 1 aerial ladder truck |
| | C | 20 chemical protection suits (no CO2 or vehicles) | | |
| | D | 1 ambulance | 1 ambulance | |
| | E | 1 pumper truck | 1 pumper truck | |
| | F | 1 police cruiser | | |
| | G | 2200 lb CO2 (no vehicles) | | |
| | H | 1 coastal patrol boat | 1 coastal patrol boat | |
| | I | 1 police cruiser | | |
| | J | 1 aerial ladder truck (no driver) | | |
| | K | 10 chemical protection suits w/breathing protection (no vehicles) | | |
| | L | 1 bus for 60 people (no driver) | | |
| | M | 4 loaded gravel trucks | | |
| | N | 20 police officers (no vehicles) | | |
| | O | 1500 ft oil booms (no transport) | | |
| | P | 12 medical personnel (no vehicles) | | |
| | Q | 1 helicopter | | |

**Table 10.2    *Concluded***

| Case | Sites | Resources | | |
|------|-------|-----------|---|---|
| | | A | B | C |
| 2 | A | 1 ambulance | 1 ambulance | |
| | B | 1 pumper truck | 1 pumper truck | 1 aerial ladder truck |
| | C | 20 chemical protection suits (no vehicles) | | |
| | D | 1 police cruiser | | |
| | E | 1 pumper truck | 1 pumper truck | 1 aerial ladder truck |
| | F | 1 ambulance | 1 ambulance | |
| | G | 1 police cruiser | | |
| | H | 1 coastal patrol boat | | |
| | I | 1 coastal patrol boat | | |
| | J | 10 chemical protection suits (no vehicles) | | |
| | K | 20 police officers | 1 police cruiser | |
| | L | 1 supermarket | | |
| | M | 4 medical doctors (no vehicles) | 10 nurses (no vehicles) | |
| | N | 1 school with gym | | |
| | P | 1 bus for 80 people (no driver) | | |
| | Q | 1 bus for 40 people (no driver) | | |
| | R | 4 medical doctors (no vehicles) | 10 nurses (no vehicles) | |
| | S | 1 office building | | |

The cases are summarized in Table 10.3. Case 2 was more severe than Case 1, since the event affected more people, and more goals needed to be met.

*Task*

The group's task in each case was to allocate resources to the incident location in order to meet response goals. In both Case 1 and Case 2, the group was told to account for decision execution time within a total time budget of 50 minutes. Non-CO

**Table 10.3  Comparison of cases 1 and 2**

| Element | Case 1 | Case 2 |
|---|---|---|
| Sites | 18 | 19 |
| Resources | 23 | 27 |
| Goals | G1–G4 | G1–G5 |

participants had to authorize the use of their resources and to specify the goals associated with that use (e.g., use of a pumper truck for the goal of fighting the fire). The CO then recorded the decisions of the group via the computer interface.

Because participants had to account for execution times, resources became infeasible if they were not used in time. Figures 10.3 and 10.4 show the number of feasible courses of action over time for each case. In both cases, the number of feasible alternatives diminishes rapidly following the onset of the case. Event severity was constant. As a result, problem difficulty increased over time since fewer resources were available to meet response goals.

Upon completion of the experiment, participants rated the quality of their solutions with respect to the overall goals of the response. Independent external judges later rated the quality of each group's solutions with respect to each goal (responses were then averaged across each goal and judge).

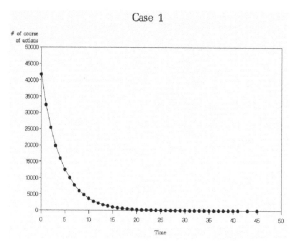

Case 1

**Figure 10.3  Feasible courses of action in case 1**

## Analytic Methods

*Data Preparation*

Data were produced from the audio/video recordings and computer log files. Conversations among group members were professionally transcribed to show (i) *time* (mm:ss) of onset of each conversational turn,

Case 2

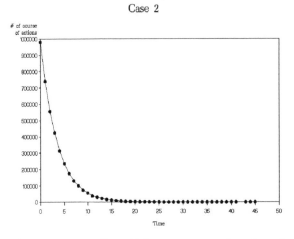

**Figure 10.4    Feasible courses of action in case 2**

(ii) *role* of the speaker, and (iii) *content* of turn (see Table 10.4 for an excerpt).

A    illustrative record    from    the computer    log    is shown in Table 10.5. At 07:52, FD sought information    about resources at site *E*. At 14:50, MO made a    recommendation that    resources    *Ea* and *Eb* each be used in achieving goal 2. At 29:22, CO submitted the decision to send resources *Ea* and *Eb* to the incident location (*Z*) to help achieve goal 2.

**Table 10.4    Excerpt of conversation transcript**

| Time | Role | Content |
| --- | --- | --- |
| 10:54 | FD | I want to send E to Z, the two closest pumpers |
| 10:59 | MO | Ea and Eb |
| 11:06 | CO | Well, what we have are two pumpers at E |
| 11:15 | PD | What kind of goals that may be? |
| 11:18 | CA | I think it is control of fire |

**Table 10.5    Sample records from log file**

| Time | Participant | Event |
| --- | --- | --- |
| 07:52 | FD | E |
| 14:50 | FD | Ea:0100 |
| 14:50 | FD | Eb:0100 |
| 29:22 | CO | Ea, Z:0100 |
| 29:22 | CO | Eb, Z:0100 |

## Data Encoding

The transcripts and log files were encoded to enable measurement of the four main constructs in the study:

- Considered course of action (CCA): a CCA is an ordered sequence of activities involving at least one resource and at least two sites. Resources in each CCA were identified, along with the sequence in which they were used and any corresponding goals. The two CCAs at 10:54 in Table 10.4 are encoded as (Ea, Z:0100) and (Eb, Z:0100), since both refer to goal 1.
- Recommendation (REC): a REC is an entry in the log file by a non-CO participant in which a resource is associated with a goal. The REC at 14:50 in Table 10.5 is encoded as (Ea:0100).
- Decision (DEC): a DEC is a course of action as recorded in the log file by the CO. The DEC at 29:22 in Table 10.5 is encoded as (Ea, Z:0100).
- Decision quality: as mentioned previously, decision quality was assessed by judges and participants.

A coding instrument was developed for identifying CCAs from the transcripts and was found to be reliable (Cohen's $\kappa$ = 0.78).

## Results

The values of the main study variables are first briefly reviewed, followed by presentation of the results for the six research questions. All statistical tests were performed at $\alpha = 0.05$.

As shown in Table 10.6, more considered courses of action (CCAs) were made and more goals associated with those CCAs in Case 2 (C2) than in Case 1 (C1). The number of resources (res) and goals per CCA was approximately equal in both cases.

Both the number of RECs and the number of goals associated with those RECs were similar across the two cases, as was the number of RECs per participant (P), and the number of goals per REC.

The number of decisions (DECs) in both cases was approximately equal, though more goals were associated with

**Table 10.6    Total, mean (std. dev.) for main measures**

| Case | 1 | 2 | | 1 | 2 | |
|------|---|---|---|---|---|---|
| CCA | 93 | 111 | CCAs | 2.43 (0.56) | 2.30 (0.70) | res/CCA |
| | 19 | 26 | goals | 0.20 (0.52) | 0.23 (0.59) | goals/CCA |
| REC | 12 | 10 | RECs | 3.00 (2.71) | 2.50 (1.73) | RECs/P |
| | 20 | 17 | goals | 1.67 (0.78) | 1.70 (0.95) | goals/REC |
| DEC | 10 | 13 | DECs | 2.70 (0.67) | 2.31 (0.85) | res/DEC |
| | 16 | 25 | goals | 1.60 (0.70) | 1.92 (0.95) | goals/DEC |
| J Eval | – | – | – | 3.75 (1.04) | 3.20 (1.64) | 1=agree, 7=disagree |
| P Eval | – | – | – | 2.60 (1.95) | 3.20 (1.79) | 1=agree, 7=disagree |

those DECs in Case 2 than in case 1. Resources and goals per decision were also approximately equal. The number of DECs in both cases represents an order of magnitude reduction from the number of CCAs.

Solutions were rated on a seven-point scale, where a 1 indicated agreement with the statement that the goal(s) had been attained, and a 7 indicating disagreement. Judges (J) and participants (P) ratings for quality of Case 1 and Case 2 DECs were not appreciably different, though it should be noted that judges' ratings for Case 1 solutions were more clearly lower than those of participants'.

In answer to *Question 1*, effort devoted to solution assembly does not appear to vary with respect to time between the low and high severity conditions. To answer the question, Figures 10.5 and 10.6 were compared, and two time series analyses conducted, as follows.

Figures 10.5 and 10.6 show the frequency with which CCAs and DECs of various sizes were observed in Case 1 and Case 2, respectively, as tallied at three-minute intervals. For example, in the first interval of Case 1, each of the six observed CCAs involved two resources (no other CCAs were observed, and no decisions

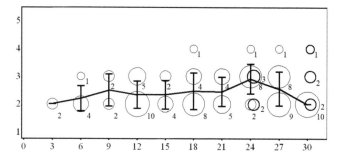

**Figure 10.5** Frequency of occurrence of CCAs and DECs of sizes two through five in case 1

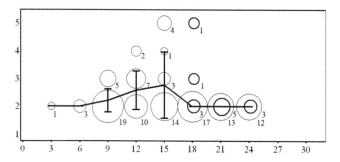

**Figure 10.6** Frequency of occurrence of CCAs and DECs of sizes two through five in case 2

were made during that interval). DECs are shown in bold. The mean and 95 per cent confidence interval for the mean number of resources in CCAs are also shown.

Level of effort is measured via the mean number of resources included in CCAs and DECs in each interval, since larger CCAs and DECs are expected to require more processing to assemble than smaller ones.

In Case 1 (see Figure 10.5), CCAs were generated through all intervals, with an approximately even split between CCAs with two and three resources, and some CCAs having four resources. There was no evidence of temporal dependency in level of effort, suggesting that – contrary to expectation – effort did not vary systematically over time. Decisions were taken in intervals eight and ten, and contained two to four resources. Timing of the initial decision was between 21 and 24 minutes, far beyond the point of greatest inflection in Figure 10.3 – though it should also be emphasized that smaller disruptions (i.e., instances of resources becoming infeasible) occur throughout most of the session. The relative constancy of effort throughout the session suggests that the buffering capacity of the group was sufficient to address the disruption indicated at the point of inflection in Figure 10.3.

In Case 2, CCAs were generated in intervals six through eight, and included two to five resources. There was no evidence of temporal dependency in the data, again contrary to expectation. Finally, decisions were taken in intervals six through eight, and included two or five resources. Thus, timing of the initial decision under the higher severity condition was at 18 minutes – beyond the point of greatest inflection in Figure 10.4 but not as far as in Case 1. As in Case 1, then, the buffering capacity of the group appears to have been sufficient to absorb or address the disruptions encountered in responding to the event.

In summary for *Question 1*, then, level of effort did not vary with time for either case, nor did level of effort increase along with level of severity.

In answer to *Question 2*, recommended use of resources deviated from convention, though level of severity does not appear to have impacted the extent of deviation. The question was answered by first identifying conventional *resource:goal* pairings, then tabulating the number of times resources of the various services were associated in RECs with the goals in each case, as shown in Table 10.7 (conventional resource:goal pairings are shown in bold). Conventional resource:goal pairings were developed before the analysis was undertaken using guidelines taken from contemporary sources (US Dept. of Transportation, 2000). For example, CA resources were associated with Goal 2 (there are no conventional goals assigned with alternative resources).

Since the number of available resources in the cases was unequal, a test was performed on the null hypothesis that the percentage of observed unconventional resource:goal mappings would be near zero (0.05) in each case ($n = 16$ in case 1, $n = 20$ in Case 2). The percentage is 37.5 per cent in case 1 and 35.0 per cent in Case 2, leading to respective *p*-values of less than 0.001 each. Consistent with expectation, there is evidence of deviation from convention in the use of resources. However, contrary to expectation, there is no evidence of the impact of level of severity, and consequently no evidence of the impact of severity on group flexibility.

In answer to *Question 3*, CCAs explain RECs only in the low severity condition (Case 1). To answer the question, a logistic regression model was fit to the data, with inclusion of resource in a REC as the dichotomous response variable, and number of times the resource was included in CCAs as the predictor variable. For Case 1 ($n$ = 23 resources), estimated parameter values in the model were significantly non-zero at $p$ <0.10, suggesting that the likelihood of a resource being included in a REC increased with the number of times the resource was included in CCAs. For Case 2 ($n$ = 27 resources), no estimated parameters values were significant, suggesting that a resource's inclusion in a REC was not affected by the number of times the resource was included in CCAs. A strong relationship between the consideration set (i.e., the boundary) would suggest high tolerance for CCAs when formulating RECs, which is clearly not the case here.

In answer to *Question 4*, the size of CCAs (as reflected in the number of resources used) does not mirror that of decisions made. To answer the question, two empirical distributions were compared. The first distribution estimated the probability of $i$ resources being included in a CCA; the other estimating the probability of $i$ resources being included in a DEC ($i=1,...$maximum number of resources in the largest CCA or DEC, whichever was greater). An exact $\chi^2$ goodness-of-fit test (Conover, 1999) was performed for each case, using a data table that represented the frequency with which $i$ resources were found in CCAs and

**Table 10.7    Resource:goal combinations for RECs**

| Case | | 1 | | | | | 2 | | | |
|------|---|---|---|---|---|---|---|---|---|---|
| Goal Role | 1 | 2 | 3 | 4 | 1 | 2 | 3 | 4 | 5 |
| CA | 0 | 2 | 2 | 1 | 1 | 0 | 0 | 0 | 0 |
| FD | 0 | 2 | 0 | 2 | 6 | 0 | 0 | 5 | 1 |
| MO | 0 | 0 | 0 | 6 | 0 | 0 | 0 | 2 | 0 |
| PD | 3 | 5 | 3 | 3 | 16 | 0 | 4 | 7 | 2 |
| AR | 1 | 3 | 0 | 0 | 5 | 0 | 1 | 5 | 1 |

DECs (*d.f.* = 2 in case 1 and 3 in Case 2). For example, in Case 2, four CCAs and one DEC contained five resources (see Figure 10.6). The null hypothesis of no association between rows and columns was retained for both cases, suggesting that the number of resources found in a CCA was independent of the number of resources found in a DEC. Knowing the size of CCAs therefore provides no indication about the size of actual decisions. As with the relationship between CCAs and RECs, then, the degree of tolerance from RECs to DECs is therefore low. In an example to follow (below), the question of tolerance is revisited more subjectively, using data drawn from one of the episodes in one of the two cases.

In answer to *Question 5*, the number of CCAs had no appreciable impact on decision quality as assessed by judges or participants. As shown in Table 10.6, more CCAs were considered in Case 2 than in Case 1. While judges' ratings are higher for Case 2 than Case 1, the difference is minimal. The direction of change in participants' ratings is contrary to expectation, though again the values are not appreciably different. With respect to margin, then, it may be said that performance was uniform relative to the boundary implied by CCAs, nor did it vary significantly based on event severity.

To provide some additional detail regarding the cognitive processes that underlaid this performance – particularly in relation to the factors of flexibility and tolerance – one episode from Case 2 is now discussed in detail.

## The Empty Bus

Recall that Phase 1 of each case required the group only to plan the response to the current event situation. Phase 2 of each case began with a report that the situation had escalated and that none of the orders for dispatching resources to the incident location had been executed. Moreover, certain resources which had been available in Phase 1 were not available in Phase 2, while other resources had been added. The added resources were so-called alternative resources: that is, resources that were somewhat unconventional for the incident, such as gravel trucks and helicopters, but that were available for use by the group.

One of the alternative resources available to the group in Phase 2 of Case 1 was a school bus with capacity for 60 people. The discussions and actions about the considered and actual use of this bus show how the group sought to link goals for the response to available materials through the generation of improvised plans, thus providing additional information on group *flexibility*. The episode of the empty bus also provides further detail on performance at boundaries of knowledge, and therefore on *margin*. We first briefly outline the discussions and actions of the group up until they began to consider how the bus might be used (all times are given as mm:ss from the time of onset of the first phase of the case).

Phase 2 begins at 37:53.[2] For approximately the first five minutes, the group exchanges information about the case and the new situation, occasionally offering considered courses of action (CCAs), but not making any decisions. At 42:43, the CO states that "The main priority now is the people that are dead, are dead." PD asks at 43:08 if personnel on the scene were "working on it" (i.e., trying to treat people or remove the dead), then at 43:11 offers "Because I have a bus that will handle 60 people. How many [injured] have we got?" The group then spends about two minutes discussing and clarifying the new situation.

The next interval of discussion (00:45:11–1:02:34) is devoted to solving the sub-problem of providing treatment to the injured. The MO mentions that he has an "overload" of medical personnel and only four ambulances. The MO then recommends that "somebody could go by L, pick up the bus, take the bus to P [where the medical personnel are located], pick up the people at P." A third option – the helicopter – is floated by CO: "Will the helicopter be faster, Jim?" (Jim is MO). PD then says he has "got a priority," meaning a problem that is high priority: he wants to send ten officers to the incident location for crowd control (i.e., G4, treatment of injured persons). MO asks "How many people can ride [in] the helicopter?" This option goes unresolved, and the group continues solving other sub-problems and discussing constraints.

It is worth noting that, during this interval, the question of how to achieve G4 is addressed from various perspectives. CCAs such

---

2     This corresponds to time = 0 in Figure 10.6.

as those above are proposed, but there is also discussion about the goal itself, as well as about how other CCAs might contribute to G4 and one or more other goals. Indeed, the generation of CCAs seems to have a priming effect in which, once a CCA is proposed, the group seeks to use it as the building block for other CCAs.

An option is considered at 46:27 to transport the medical personnel to the scene by ambulance. The police personnel are then left without this option. The available patrol cars are sufficient to transport only half of the ten personnel PD wants to send. The CO at 47:25 states "The more you can handle yourself, handle yourself," a comment which is fielded by MO, not PD, who says he didn't know that he could fit 12 people in his ambulances. He says he still needs personnel, leading MO to comment that he has "no means to transport injured now" (48:01), since presumably the arrival of the ambulances will be delayed by sending them to pick up medical personnel. At 48:16 MO says "Helicopter, I am using it now," but CO says "No." CA notices the exchange and asks "Who owns the helicopter?", to which MO replies "Anybody." MO says very little, while from 48:25 until 50:38 the group works on how to get firefighting equipment to the scene (they also consider CCAs for combating the oil slick).

At 52:24 the group returns to the question of how to transport police personnel. They decide to use the police cruisers to pick up the chemical protection suits. Rather abruptly (see Table 10.8), the bus is brought back into discussion.

Based on the statement at 53:23, PD does not know who "got" (i.e., picked up) the bus. MO confirms that the bus does not have its own "people" (i.e., people who can drive it). PD then outlines a complete course of action that would involve taking a police cruiser (*Fa*) to L, using the driver of the cruiser to pick up the bus, going to N, then continuing to the incident location (Z). Police resources would therefore serve as the lynchpin for two PD resources and an alternative resource.

Following a question and comment by MO, PD reconsiders the decision to send the bus to Z (55:03), as shown in Table 10.9. FD offers a possible contingency plan (i.e., using the fire engines) if the ambulances are insufficient for transporting the victims (55:10). CA then offers a second contingency plan, which is using

**Table 10.8    First episode**

| Time | Role | Statement |
|---|---|---|
| 05317 | PD | At *L*. |
| 05318 | CO | To *Z*. |
| 05320 | PD | What was at *L*? |
| 05321 | CA | I never burnt up a ship before. |
| 05322 | FD | Bus. |
| 05323 | PD | Who got the bus? Am I getting the bus too? |
| 05326 | MO | Yeah, I... |
| 05327 | PD | So if I take the bus... If I send *Fa*... |
| 05330 | CO | Does the bus have its own people? |
| 05332 | MO | No. |
| 05333 | PD | It doesn't have a driver. |
| 05334 | CO | Does the bus have a driver? |
| 05335 | MO | No. |
| 05336 | PD | So what I am going to have to do is take *Fa* and leave it to get *La* to go to *Na* and then up to *Z*, right? How many steps can one person do? |
| 05343 | CO | As many as you want. |

the bus for victim transfer (55:16). Clearly, a lot is being asked of this bus.

PD raises a concern that the wind direction might bring toxic fumes into the path of the bus. They then consider alternative methods for getting emergency personnel in and victims out. The course of action for the bus is entered at 1:02:34 as *Fa>La>Na>Z* (i.e., cruiser to bus to pick up suits, then to incident location), the goals being G1 (control of access to incident location) and G4 (treatment of injured persons).

**Resilience Factors Revisited**

This approximately 17-minute-long episode shows how concern with satisfying a single goal (patient treatment) through the use of a single resource – an empty bus – evolved through many complex additions and deletions into a single course of action involving

**Table 10.9    Second episode**

| Time | Role | Statement |
|------|------|-----------|
| 05450 | MO | What you are going to do with the bus? |
| 05452 | PD | I have got... What I am doing is I am taking *Fa* up to *L* which has the bus, right? And then I am going to go back to *N* and get my police officers and then, um... |
| 05502 | MO | Go to *Z*. |
| 05503 | PD | Do I need to go to *Z* at that point? I don't need to get your ambulance personnel, so I am going to *Z* at that point. |
| 05510 | FD | You can always, if you got enough backboard, you can transport some patients in the fire engines. |
| 05515 | MO | Yeah if you... |
| 05516 | CA | Or on the bus. |
| 05517 | PD | Or on the bus. And that's the other reason to get both the police officers in the bus there to handle... |

three resources (drawn from two services) and one additional goal. It is only by tracing the explicit connections between CCAs (via both resources and goals, as well as occasionally through more abstract concepts) that details of divergent and convergent processes may be seen.

As stated previously, descriptive and prescriptive models of creativity typically emphasize the need for divergent thinking early in the process, followed by convergent thinking and, ultimately, decision. In the episode of the empty bus, it is perhaps accurate to say that a single seed of an idea grew and was pared down to a still complex but nonetheless manageable structure. Divergent thinking in this process may be seen in the various approaches taken by the group in seeking to exploit this resource (e.g., by moving personnel, victims), as well as in the various candidate courses of action that seem to have been generated in an attempt to utilize the resource. Convergent thinking may be seen in how goals and contextual and material constraints were considered. In other words, invocation of goals and constraints helped to narrow the set of CCAs. Clearly, in this episode, divergent and convergent thinking are tightly coupled, with the results of one process feeding the other.

All of the factors except cross-scale interactions may be seen in this study, and special attention has been given to two of them – flexibility/stiffness and margin – in the discussion of the empty bus. The perspective throughout this chapter has been that the resources available to the group may be cognitive, behavioural and material. With respect to *buffering capacity*, neither changes in severity within or across events appears to impact the adaptive capability of the group. Regarding *flexibility/stiffness*, the episode of the empty bus in particular illustrates the importance of examining group processes in detail in order to identify strategies for achieving flexibility. Indeed, the data encoding used in testing the research questions relative to flexibility/stiffness does begin to pry apart group processes, but tells only part of the story. Through an examination of the empty bus episode, it may be seen that the relationship between CCAs, recommendations and decisions is potentially quite fragile, subject to fissures arising from competing demands on attention, discordant preferences among group members, limited processing capacity of key personnel, or simply noise in the environment.

Regarding *tolerance* and *margin*, the boundaries of group performance may be established by external forces (such as the event itself, or by policy-level concerns), but – as has been seen here – in a more subtle way by preferences and decisions expressed by group members. Indeed, despite the best efforts of the designers of experiments to assess organizational resilience, the very construct of performance itself may be subject to redefinition by the group.

## Discussion

Emergency response organizations operate in environments that are characterized by time constraints and the frequent (and often unpredictable) occurrence of unplanned-for contingencies. To the extent that these groups can effectively marshal available cognitive, behavioural, and material resources in such environments, they are resilient. The approach taken here has been to examine the cognition and behaviour of one group in addressing two simulated emergencies that differed in their level of severity, and thus to assess their resilience. In general, the group was *resistant to* increases in severity, as well as to the smaller contingencies that occurred within each case, in the sense that their overall performance did not change with changes in severity. The results suggest a reasonable amount of *resilience* in the group. Level of severity did not have an impact on buffering capacity or flexibility. On the other hand, the results for Question 3 strongly suggest that there were no significant links between considered courses of action, recommendations, and decisions – thus suggesting low tolerance. The analysis of the empty bus shows that the processes leading to decision ought to be examined at an even more detailed level. Future work in this area may, for example, be devoted to analyzing the links between individual considered courses of action, recommendations, and decisions.

## Conclusions

Decision-making in emergency response takes place in relation to potentially high costs associated with human, material and economic losses, and a concomitant need to organize multidisciplinary groups to address highly non-routine situations. This study suggests that level of event severity had little appreciable effect on the quality of solutions generated by one group. But perhaps more importantly the methods and results demonstrate the degree of resilience among the group, as manifested in the thinking and behavioural processes that engendered their decisions. The results have offered some insights into the group processes that follow the establishment of an ERO, but also into broader questions of the progress of creative decision-making in groups. Future work in this area may benefit from the use of a wider range of levels of severity, particularly to uncover the impact of this factor on the timing of decisions.

# Chapter 11

# Restoration Through Preparation: Is it Possible? Analysis of a Low-Probability/ High-Consequence Event

Martin Nijhof
Sidney Dekker

## Introduction

Creating resilience through preparation and restoration embodies a fundamental tension. No complex, dynamic operating world is likely to be immune to this conflict on how to absorb the cognitive and coordinative demands that come with low-probability/high-consequence events. **Preparation** of practitioners for safety-critical situations is mostly organized around *a priori* control, and driven by a limited number of predictable scenarios (laid down in training and procedures for how to handle particular situations). **Restoration** of a safety-critical situation, on the other hand, often relies on practitioners departing from protocol, and on them extemporizing and importing new knowledge and skills (e.g. Weick, 1988; Orasanu et al., 2001; Dekker, 2003; Dismukes et al., 2007)

Preparation, in other words, may assume that the demands of high-consequence/low-probability events can be absorbed by matching situational symptoms with pre-fabricated scripts of coordinated action that not only appear in procedures such as the Quick Reference Handbook (QRH) (onboard flight decks) but also get rehearsed in practitioner training. These forms of preparation support the prioritization of actions and callouts in the face of time pressure and resource constraints, help assign tasks, set the pace and order for the work to be done, organize roles and calibrate mutual expectations of activity and double-checking. However, the limits of, and misplaced confidence in, such preparation have been commented on before (e.g., Suchman, 1987; Wright and McCarthy, 2003; Burian and Barshi, 2003), and the literature has shed some light on the difficulty of processes of sensemaking in demand situations that lie beyond procedural reach (Weick, 1993).

As Hollnagel (this volume) reminds us, resilience asks not what is *absent* in a system so that it becomes unsafe. Instead, it asks what capacities are *present* that can help the system manage unforeseen situations both large and small, and what can be done to enhance such adaptive capacities. In this chapter, we present an example that may illuminate the role that preparation plays in eroding or enhancing the capacity to restore a high-demand, unforeseen critical event. The event in question (oil loss on both engines of a twin-engine airliner) happened to the highly reliable Boeing 737 (it has flown in various models for over 40 years). As a contrasting case, we describe the way in which a related situation was managed on a different occasion. In the first case, restoration was brought about by improvisation. In the other case, restoration was enacted through adherence to what had been prepared through procedure and training. The distinction may serve as a meaningful commentary: how activities associated with successful restoration not only depart from those practiced and published in preparation, but how the latter could actually exacerbate the problem and possibly greatly amplify its negative consequences.

We finish by discussing how to enhance practitioners' ability to *reliably* create resilience through extemporization (as our contrast cases testify, such reliability seems absent today). Here we deliberately sustain the fundamental tension: as extemporization is in principle unpredictable, unique and not replicable, how can practitioners' adaptive capacity be enhanced through replicable preparation?

## Example 1: Low Probability/High Consequence

During the climb of an otherwise normal Boeing 737 flight from East Midlands airport, the commander was carrying out duties as the non-handling pilot (on a two-pilot flight deck, one pilot is always designated "pilot flying," or "pilot handling"; the other is "pilot monitoring," or "non-flying." These roles often get swapped on each new leg). Here, these duties included monitoring the engine instruments which initially indicated no problems. However, as the aircraft was climbing to approximately 14,000 feet (Flight Level or FL 140), the commander noted that both engine oil quantity gauges were indicating about 15 per cent of normal quantity (this is very low); the indications were also fluctuating and continuing to decrease.

As the crew was discussing this problem, both engine oil pressures also began decreasing. Normally, oil from each individual engine tank is circulated under pressure through the engine to lubricate the engine bearings and its accessory gearbox. The oil is pressurized by an engine-driven oil pump, and returned to the oil tank by an engine-driven scavenge pump, both of which should work as long as the engine works. Low oil quantity and oil pressure on both engines were confusing indications, particularly since the two oil systems are completely separate, and could have lead the commander to assume some sort of indicating error.

The aircraft was transferred to London control and, at 12:04, the commander contacted London and was cleared for further climb, to Flight Level 210. But a minute later, the oil quantities continued to decrease and the oil pressures were now below 13 psi. Pilots are not required to remember numeric indications as too low or too high. Instead, the presentation colour changes from white to amber (caution) to red (limit). The commander informed London that he had engine problems and wished to return to East Midlands Airport.

The controller immediately cleared the aircraft to level off at FL 180 and to turn left. By now, both engine oil quantities were indicating zero and the oil pressures were still decreasing. The crew informed Air Traffic Control (ATC) of the deteriorating situation and declared an emergency. As the aircraft was turning left onto the designated heading, the ATC controller informed the crew that Luton Airport was the closest diversion, and the crew agreed it was a better option. Then they commenced a descent, after having been level at FL 180 for only 17 seconds. With the emergency declared, the commander assumed the role of handling pilot, and the first officer was allocated the tasks of re-programming the flight management computer to navigate to Luton and execute the approach and landing there. The first officer also operated the radio.

During the descent and approach to Luton, the ATC controller assisted the crew by minimizing transmissions but passed them all essential information such as continuous descent clearances and weather information for Luton. Within the aircraft, the flight crew were working together to ensure that the aircraft was configured for the approach to Luton. The commander was using idle power and had made the decision that they would be making a Flap 15 landing. He decided on this flap setting because he was very conscious of the need to minimize power requirements due to the engine oil situation (a normal landing occurs with flaps at 30 or 40 degrees, which offers lower landing speeds but creates more drag which in turn requires greater engine thrust to overcome); both low oil pressure lights had illuminated 40 seconds after leaving Flight Level 180.

The commander had called the senior cabin attendant to the flight deck and fully briefed her on the situation, including the fact that they were making an immediate diversion. In turn, the senior cabin attendant briefed the cabin crew and the cabin was quickly prepared for landing. The commander had already briefed the passengers that he was diverting to Luton. The approach was direct and uneventful, and the crew was given priority and all requested assistance by London and Luton ATC agencies. The aircraft touched down on Runway 26 at Luton at 12:14.

Immediately after landing, the commander selected idle reverse power and used maximum automatic braking. With the landing roll under control and after transferring electrical power to the Auxiliary Power Unit, the first officer (on instruction from the commander) closed down both engines while still rolling on the runway. The aircraft came to a stop on the runway and was quickly attended by the Luton Airport Rescue and Fire Fighting Service (RFFS). Verbal contact with them was quickly achieved through the open flight deck window and the commander took the decision not to carry out an emergency evacuation. The RFFS remained on the scene and the passengers began disembarking through the front left door using the aircraft air stairs. Once the commander had disembarked, he noted that the cowls of both engines, and the underside of both wings and flaps immediately behind the engines, were covered in engine oil.

The aircraft had been subject to a maintenance inspection on both engines during the night prior to the incident flight. The inspection required removal of the High Pressure rotor drive covers, one on each engine. After the inspection had been completed, the High Pressure rotor drive covers had not been refitted. This resulted in the loss of almost all of the oil from both engines during the flight (AAIB, 1996).

**The Procedure for Low Engine Oil**

There is always at least one paper Quick Reference Handbook (QRH) on the flight deck of a 737. They are readily accessible to the flight crew, and the (supposedly) applicable procedure can be found efficiently through a contents list that helps crews match observed symptoms with what they should do. There is no procedure in it for low oil quantity, as an airplane is not supposed to even be dispatched with an oil quantity that is too low. Also, low oil quantities will almost invariably be accompanied by low oil pressure. Assessment of the situation and subsequent actions in the event of low oil pressure are now "proceduralized" and trained as follows:

> Accomplish this procedure when the engine oil pressure is below 26 psi or when the amber LOW OIL PRESSURE light is illuminated. The amber LOW OIL PRESSURE light illuminates at a pressure below 13 psi.
>
> If engine oil pressure is in the amber band with take-off thrust set:
>
> Do not take off
>
> If engine oil pressure is at or below the red line:
>
> Accomplish the ENGINE FAILURE / SHUTDOWN CHECKLIST

Here, the QRH and pilot training assume that low oil pressure that necessitates an engine shutdown will happen on only one engine at a time. There is no procedural or training provision for this happening on two engines simultaneously (which would naturally prompt a crew to first suspect an indicator failure, especially if the engines keep running despite the indications). In the situation the crew members were in, there was no advantage to them in closing down either or both engines.

## Example 2: Low Probability/High Consequence

This is a contrasting case which did involve an indicator error. A crew on the same type of aircraft had just reached the top of a climb at 35,000 ft when they observed that the engine oil pressure indicator of the right engine was showing in the red area. The corresponding LOW OIL PRESSURE light, however, was not illuminated. The crew did follow the non-normal checklist procedure for engine low oil pressure, and shut down the right-hand engine. After contacting maintenance, the aircraft returned to its departure field and made an uneventful approach and landing. Once on the ground, maintenance people replaced the right-hand oil pressure indicator and conducted a satisfactory engine test run.

One could argue that this was procedure-following (resulting in a shutdown of an engine that had no problems) in the face of a situation that, in retrospect, called for more subtle interpretation and adaptation of the response. Interestingly, such nuance *had* been present in the previous procedure, which had been changed by the aircraft operator in a drive to standardize its checklists:

LOW OIL PRESSURE light illuminates

–    Perform ENGINE FAILURE AND SHUTDOWN procedure

LOW OIL PRESSURE light does NOT illuminate

–    Oil parameters ....................................MONITOR

Procedure completed.

This choice moment had been removed, and with it a guide to successful adaptation in a subtly different situation (i.e., an indicator failure). Of course, in this case only one engine (or its indication) was affected, and shutting it down (although, in hindsight, unnecessarily) is not as critical as having two ailing engines. This case also shows that the tension between centralization (determining solution strategies beforehand) and decentralization (leaving solution strategies to those on the front-line) affects those who develop the procedures as much as those who do operational work with them. Detailed specification, as was removed in this case, could still have helped in a narrow set of circumstances and allowed people to restore problems with help from a QRH, but clashed with other goals, such as the standardization and broader applicability of the procedural guidance left in place.

## The Procedure and the Training

Equipment vendors of course understand that there is a distance between preparation through procedures and training, and restoration from real problems, particularly multiple nested problems:

> While every attempt is made to provide needed non-normal checklists, it is not possible to develop checklists for all conceivable situations, especially those involving multiple failures. In some unrelated multiple failure situations, the flight crew may ... exercise judgment to determine the safest course of action. The captain must assess the situation and use good judgment to determine the safest course of action.
>
> *(Boeing 737 Flight Crew Operations Manual Quick Reference Handbook, 2007, p. CI 2.2)*

What such "good judgment" is, however, is left implicit (possibly because good judgment is entirely context-dependent), but there is also no mention of any particular way in which an airline could help prepare crews to exercise such "good judgment" in situations not covered by prepared scripts or training. Whereas aviation offers ample lip service to the importance of pilot judgment (such as in Crew Resource Management training and in the QRH above), there is no stronger arbiter on whether that judgment was good or bad than the outcome of the situation. After all (see also Woods and Shattuck, 2000):

- If rote rule following persists in the face of cues that (in hindsight) suggest procedures should be adapted, this may lead to unsafe outcomes (see Example 2: safety margins reduced considerably and, in retrospect, unnecessarily). People can get blamed for

their inflexibility, their application of rules without sensitivity
to context.

- If adaptations to unanticipated conditions are attempted without
complete knowledge of circumstance or certainty of outcome,
unsafe results may occur too. In this case, people get blamed for
their deviations; their non-adherence.

In other words, people can fail to adapt, or attempt adaptations
that may fail. Rule following can become a desynchronized and
increasingly irrelevant or even dangerous activity, decoupled from
how events and breakdowns are really unfolding and multiplying
throughout a system. But potential adaptations, and the ability to
project their potential for success, are not necessarily supported
by preparation in training or overall professional indoctrination.
Civil aviation tends to emphasize a model where respect for
procedures will most likely enhance safety (e.g., Lautman and
Gallimore, 1987), but any situation not contemplated in the design
and operation of the airplane, and therefore not covered in the
QRH, is reserved for the pilot's own "good judgment."

## Conclusion

Letting people adapt without adequate skill or preparation for
such adaptation can increase the risk of failed adaptations. A
better way out of the double bind is to develop people's skill at
adapting. In a way, of course, aviation has long since attempted
to do exactly this through a constantly evolving program of Crew
Resource Management (CRM) training. It offers teaching for the
kinds of generic skills necessary for managing low-probability/
high-consequence events. There are moves all over the industry
to bring such skills (and the concepts and language in which to
discuss them) closer to the actual piloting task (e.g., by integrating
CRM concepts in simulator training, and by trying to debrief
and reflect on pilot performance during, e.g., line checks using
those concepts). Piloting skills, after all, have traditionally been
taught in an extremely instrumental way, anchored from the very
first moments to the actual control of mechanical and computer
interfaces in a cockpit (or its simile).

Opportunities have been scarce during pilot preparation for a build-up and rehearsal of general competencies such as information management (sorting, prioritizing, explicit goal statements), leadership and communication (flexibility, building up shared mental models), proactive strategies and analysis, and assessment of any intervention effects and revision of plans. And still, it is often up to individual instructors to come up with scenarios that challenge or rule out routine solutions. Their time to really develop such scenarios and get crews to try different solutions paths is severely constrained by the shrunken training footprint available to airlines today, and the overflowing package of standard skills that need to be trained instead. To become complete and meaningful, such excursions into non-written, non-QRH problem restoration require the development of judgment about local conditions and the opportunities and risks they present, as well as an awareness of larger goals and constraints that operate on the situation. Development of this skill could be construed, to paraphrase Rochlin (1999), as planning for surprise. Indeed, as Rochlin (1999: 1549) observed, the culture of safety in high-reliability organizations anticipates and plans for possible failures in "the continuing expectation of future surprise."

# Chapter 12

# Understanding and Contributing to Resilient Work Systems[1]

Emilie M. Roth
Jordan Multer
Ronald Scott

Resilience engineering attempts to understand and contribute to the design of resilient work systems. For a work system to achieve and maintain resilience, the practitioners involved must be able to: (1) actively anticipate potential failure paths and adapt work practices so as to avoid them; and (2) continuously evolve work practices so as to maintain operation within safe bounds while meeting efficiency/productivity objectives. We use examples drawn from the domains of railroad operations and military command and control to illustrate these points. We argue that in order to foster resilient operations, work-centred support system designs need to accommodate both formal and informal work practices, and to provide mechanisms to enable systems to evolve to keep pace with changes in work practices that will inevitably arise in response to changing work demands.

## Introduction

Resilience engineering has emerged as a field that attempts to understand and contribute to the design of resilient work systems.

---

1       Portions of this work were supported by the US Federal Railroad Administration Human Factors Program. The WCSS and evolvable systems research was supported by the Air Force Research Laboratory Human Effectiveness Directorate. The views of the authors do not purport to reflect the position of the Federal Railroad Administration, the US Department of Transportation, or the Air Force Research Laboratory Human Effectiveness Directorate.

One of the hallmarks of a resilient work system is the ability to continue to function effectively in the face of escalating demands and unforeseen circumstances (Hollnagel, Woods, and Leveson, 2006; Nemeth et al., 2008). This requires the ability to anticipate the ways in which things can go wrong, to develop strategies that are resistant to failure, and to adjust tasks and activities so as to maintain safety margins (Nemeth et al., 2008).

This chapter presents two case studies to illustrate that for a work system to achieve and maintain resilience, the practitioners involved must be able to: (1) actively anticipate potential failure paths and adapt practices so as to avoid them; and (2) continuously evolve work practices so as to maintain operation within safe bounds while meeting efficiency/productivity objectives.

System designers need to take these precepts into account when attempting to develop effective, work-centred, support systems (Roth, Scott, Deutsch et al., 2006). Workers actively contribute to system resilience through both formal work practices that are governed by documented rules and procedures and informal work practices that emerge in response to work demands. To foster resilience, support systems need to accommodate both formal and informal work practices and to evolve to keep pace with changes in work practices that will inevitably arise in response to changing work demands.

The next section presents a case study drawn from railroad operations that illustrates how informal work practices can contribute to resilience – enabling railroad workers to anticipate and avoid accident situations (Roth, Multer, and Raslear, 2006). The case study reveals the importance of understanding how features of the current environment contribute to resilience, as a foundation for developing next-generation technologies, so as to avoid inadvertently disrupting these strategies and degrading resilience.

This is followed by a case study drawn from a military command and control application. This study reinforces the value of informal practices for maintaining safe and efficient operations. In addition it illustrates how resilience depends on the ability of domain practitioners to adapt and evolve work practices in the face of change. It argues for the importance of developing "evolvable" work-centred support systems that enable domain practitioners to extend and adapt software tools so as to maintain resilience in the face of rapid change in work demands (Roth, Scott, Deutsch et al., 2006).

## The Role of Informal Practices in Contributing to Work System Resilience

The first case study illustrates the contribution of informal proactive practices to resilience in the context of railroad operations. The study, which is more fully described in Roth, Multer, and Raslear (2006), examined the role of informal strategies for developing shared situation awareness within a distributed work system consisting of roadway workers, train crews, and railroad dispatchers. These strategies emerged informally and contributed to system resilience by enabling individuals to anticipate and avoid miscommunications with safety consequences.

Railroad operations require coordination among individuals widely distributed in space. This includes train crews that operate trains across multiple territories that are controlled by different dispatchers; roadway workers that maintain the tracks, signals, and related infrastructure; and dispatchers that manage track usage, allocating time on the track to different trains and roadway worker activity as required. These individuals rely heavily on analogue radio communication to maintain awareness of each other's location, coordinate work, and maintain safe operations (Roth, Malsch, and Multer, 2001; Roth, Malsch, Multer, and Coplen, 1999; Roth, Multer, and Raslear, 2006; Roth and Patterson, 2005). For example, dispatchers use radio communication to provide permissions to roadway workers to occupy particular portions of track for specific time periods. Radio communication is also used to control train movement in "dark territory" that do not have signals for directing the movement of trains. In dark territory, the train crew is given permission to move across specific blocks of track. When they approach the end of the portion of track for which they have permission, they must call the dispatcher to indicate their location and request permission to move across the next portion of track.

Most communication among railroad workers is governed by formal communication protocols. For example, when a dispatcher issues an authority to occupy track to a roadway worker or locomotive engineer, the individual is required to repeat back the information to insure it was properly received and understood. As is discussed below, railroad workers have developed additional informal communication practices not required by formal rules or regulations, which contribute additional layers of system resilience.

A goal of the research was to understand the factors that affect roadway worker safety in today's environment so as to anticipate the likely impacts of emerging technologies on roadway workers and to provide guidance for design and introduction of the technologies. The railroad industry is developing a number of new technologies that will affect roadway workers. This includes new forms of Positive Train Control (PTC) that are designed to protect roadway workers and prevent train-to-train collisions by providing backup warnings and, if necessary, automatically stopping trains that exceed speed restrictions or enter track segments for which they are not authorized. A second, related, technology that is emerging is digital communications that are intended to replace analogue radio. Current analogue radio has a number of unique features with positive and negative aspects. Most salient is the "party-line" aspect of analogue radio communication. Communication over analogue radio can be overheard by anyone within reach of the analogue radio signal. One of the consequences is that the radio channels can become very crowded making it difficult for messages to get through. This, combined with the fact that radio signals can become easily degraded (e.g., due to signal interference or environmental conditions), has led railroads to investigate alternatives. One example is portable digital communication devices intended to enable roadway workers to communicate more reliably.

The goal of the research was to anticipate and help to shape the impact of these new technologies on railroad operations so as to enhance overall system resilience. This includes identifying ways technology can be used to more effectively support cognitive and collaborative activities that contribute to safety, as well as avoiding introduction of technology changes that could disrupt effective strategies without providing alternative ways of achieving the same safety objectives.

The study combined interviews with field observations. Site visits and interviews were conducted at five locations in the United States and included passenger and freight rail operations. Among the most notable findings of the study is that railroad workers have developed a variety of informal practices that in combination provide multiple layers of resilience, contributing to safe operations. In particular, observations and interviews revealed a number of proactive, cooperative practices that contributed to roadway worker safety. The practices were informal in the sense that they were not dictated by formal rules, procedures, or regulations. They were proactive in the sense that they were voluntary actions intended to prevent problems from arising. They were cooperative in the sense that they were specifically intended to support the activity and safety of others.

A complete description of the study methods and results is provided in Roth, Multer, and Raslear (2006). A summary of the key findings related to the contribution of informal worker activities to resilience is provided below.

*Informal Strategies Contributing to Resilience in Railroad Operations*

Roadway workers inspect, maintain, and repair railroad facilities and equipment including track, signals, communications, and electric traction systems. They may work alone or as part of a multi-person group that must coordinate their work to accomplish a common task. Some jobs require working at a particular location on the track (e.g., changing a rail, troubleshooting a malfunctioning signal). Other jobs require moving over the track, for example to inspect track.

Because the activities of roadway workers are performed on or near railroad tracks, they are at constant risk of being struck by a train or other on-track equipment. A review of a Federal Railroad Administration roadway worker fatality data set covering the period from 1986 through 2003 revealed that more than 65 per cent of fatalities were caused by a train. Thirty-four per cent occurred while working on the track on which the train was running, 17 per cent occurred while working on an adjacent track, and 16 per cent occurred while walking to or from the worksite.

One of the most notable findings of our study is that railroad workers have developed a variety of informal cooperative practices that contribute to roadway worker safety. This includes informal strategies that have been developed that serve to foster shared situation awareness among roadway workers, train crews and dispatchers, with respect to the location and activities of railroad workers in a given territory. Railroad workers have also developed "cross-checking" strategies for catching and correcting errors that may have safety implications. Table 11.1 provides examples of these informal cooperative practices. In combination, these informal strategies provide multiple layers of resilience, contributing to safe railroad operations.

One important strategy that has emerged relies on the "party-line" aspect of analogue radio communication. When railroad workers communicate over the radio their communication can be overheard by anyone listening on the appropriate channel, if they are within reach of the signal. For example, if roadway workers monitor the appropriate radio channels, they can overhear communications between dispatchers and train crews that can reveal information about the location and planned movement of the trains. Train crews, roadway workers, and dispatchers all actively worked to extract relevant information by "listening in" on radio communications directed at others.

These active listening-in processes enable individuals in the distributed organization to identify information that has a bearing

**Table 11.1    Informal cooperative strategies for maintaining shared situation awareness and enhancing on-track safety**

Roadway workers monitor radio channels to extract information about trains in their vicinity.

Dispatchers monitor communications directed at others to maintain awareness of location and activities of trains and roadway workers.

Dispatchers call to alert roadway workers of trains – particularly if they are coming at a non-usual time or unexpected direction.

Train crews call other trains to alert them of roadway workers in their vicinity.

Roadway workers call other roadway workers to alert them of trains heading in their direction.

*Source: Adapted from Roth, Multer, and Raslear, 2006.*

on achieving their own goals or on maintaining their safety. For example, roadway workers and train crews routinely exploit the "party-line" aspect of radio communication to build and maintain awareness of the location, activities, and intentions of others in their vicinity. Interviews with roadway workers and dispatchers indicate that roadway workers actively monitor radio channels on which train crews communicate in order to anticipate when trains are likely to approach them and in what direction. This is particularly important in the case of unscheduled trains that may appear at a different time or from a different direction than expected.

Listening-in strategies also enabled railroad workers to recognize situations where information in their possession was relevant to the performance or safety of others. We documented several instances where third parties overhearing conversations played an instrumental role in preventing accidents. For example, in several cases, individuals overheard conversations suggesting that someone was (erroneously) occupying track for which they did not have authority, or were about to be (erroneously) given authority to enter track that was already occupied. They immediately got on the radio to alert the parties of the potential conflicts. By intervening to prevent another party from erroneously occupying track for which someone else was authorized, they prevented a potential collision between equipment and/or people. These are examples of proactive, cooperative strategies that serve to enhance the overall safety of railroad operations. The results reinforce findings from other domains (e.g., space shuttle mission control, air traffic control, aircraft carrier operations) regarding the importance of informal listening-in strategies for supporting anticipation, contingency planning, and catching errors and recovering from them (Luff et al., 1992; Patterson et al., 1999; Rochlin et al., 1987; Smith, McCoy, and Orasanu, 2001).

In addition, to active listening-in processes, cooperative communication practices have emerged that are specifically intended to enhance safety on the track by improving shared situation awareness of the location and activities of roadway workers and trains operating in the same vicinity. Observations and interviews revealed proactive communication practices that went beyond the requirements of formal operating rules, and served to foster shared situation awareness, facilitate work and enhance on-track safety. For example, dispatchers, train crews and other roadway workers routinely alert roadway workers of trains that may be about to reach them, particularly when these trains arrive at an unexpected time or from an unexpected direction. As one dispatcher stated, "I let them know what my plan is so that they are not startled." This call is not mandated by operating rules. Similar informal communications that provide an important safety function have been observed among train crews. For example, when a train crew passes a roadway worker group working by the side of the track, the conductor may call over the radio to alert other trains passing through the territory of the presence of the roadway workers. Instances were also observed where roadway workers travelling on track cars called other groups of roadway workers that they had passed earlier to alert them to a train heading their way. In all cases the goal of the communication is to enhance awareness of the location, activities, and intentions of others operating in the same vicinity.

Interestingly, these practices are referred to as "courtesies" by the railroad workers. They are not required by the operating rules. However, these informal, cooperative practices play an important role in enhancing system resilience by enabling roadway workers to anticipate and prepare for trains heading their way that might otherwise have been unforeseen, preventing mishaps. They are part of the informal redundant "safety net" that is provided through voluntary proactive activities among railroad workers.

Figure 12.1 summarizes the role that informal information extraction and information sharing processes play in enhancing overall work system resilience. Distributed railroad workers actively work to build and maintain shared situation awareness of the location, activities, and intentions of roadway workers and trains in a given vicinity. This shared contextual knowledge in turn contributes to informal cooperative practices that enable the distributed team to anticipate and avoid problems, as well as catch and correct errors with potential safety consequences, enhancing the overall resilience of the distributed organization.

These informal practices provide concrete illustrations of cooperative information sharing and "cross-checking" activities that have been found to contribute to overall system reliability

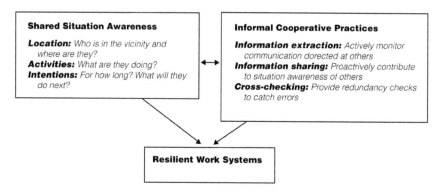

**Figure 12.1    Framework showing how informal cooperative processes across distributed teams build on and contribute to shared situation awareness and how the two combined can foster resilience contributing to higher reliability performance and increased safety**

*Source: Adapted from Roth, Multer, and Raslear, 2006.*

in other domains (Branlat, et al., 2008; Rognin, et al., 2000). The current results highlight that these information sharing and cross-checking activities are not necessarily built into the formal organizational structure, but rather emerge as informal, proactive practices intended to increase system reliability and resilience to error.

The fact that these communication practices were frequently referred to as "courtesies" highlights their optional nature, and positive contribution, to overall system resilience. They provide reinforcing evidence that informal practices of domain practitioners can contribute substantively to system resilience and safety (Hollnagel, Woods, and Leveson, 2006). This view contrasts with more traditional perspectives that emphasize the role of humans as sources of "error" that can degrade an otherwise safe system.

*Implications for New Technology Insertion*

One of the dangers of introducing new technology is that it will disrupt effective strategies that contribute to system resilience. In order to avoid inadvertently disrupting effective strategies that contribute to resilience, one must begin by understanding the cognitive and collaborative activities that currently contribute to system resilience, and the aspects of the current environment that support those activities. In the case of railroad operations, our analysis revealed a variety of informal strategies that roadway workers, locomotive engineers, and dispatchers have developed to maintain shared situation awareness of each other's location, activities, and intentions. These include active "listening-in" strategies that depend on the "party-line" aspect of radio communication. These active listening processes enabled individuals in the distributed organization to extract information that had a bearing on achieving their own goals, as well as to recognize when information in their possession needed to be shared so as to support the performance or safety of others. These findings have relevance for how new position localization and digital communication technologies might best be deployed to foster greater shared situation awareness across the distributed work system. They also point to potential pitfalls that should be avoided.

New technologies are emerging that have the potential to facilitate communication and more effectively support the cognitive and collaborative processes required for maintaining shared situation awareness among roadway workers, dispatchers, and train crews – provided they are deployed appropriately. Railroads are developing portable devices for roadway workers that integrate location-finding technologies (e.g., GPS) for more accurate location information and digital technologies for more reliable communication. These devices could be integrated with positive train control systems to enhance mutual awareness of the location, activities, and intentions of trains and roadway worker groups within the distributed organization, increasing overall safety. However, unless carefully deployed, these technologies have the potential to disrupt the strategies that practitioners currently use to maintain shared situation awareness resulting in a degradation of resilience. In particular, if digital technology were deployed so as to eliminate the party-line aspect of current radio technology, without providing an alternative means to foster shared situation awareness, it could degrade the ability of individuals to maintain awareness of each other's location, activities, and intentions.

Properly deployed, location finding and digital communication technologies have the potential to reduce the challenges associated with analogue radio communications while still providing the kind of situation awareness information that is now extracted indirectly. For example, location-finding technology makes it possible to develop graphic displays that directly show the location of roadway workers and trains in a given vicinity. The same information display could be made available to dispatchers (on a display in the dispatch centre), roadway workers (on portable graphic devices), and train crews (on a display in the locomotive cab). Thus, location information that is important for shared situation awareness, which is now extracted indirectly (e.g., by listening in to radio communications directed at others), could be obtained more directly and with lower cognitive overhead.

There is also the opportunity to preserve the virtues of the broadcast aspect of analogue radio when transitioning to digital communication technology. One of the benefits of analogue radio is that it enables the actions and intentions of parties to be broadcast to multiple individuals at once. Digital communication technology can be deployed in such a way as to preserve this positive broadcast feature. For example, a dispatcher could specify multiple parties that should receive a given audio message so as to foster shared situation awareness of the activities and intentions of trains and roadway workers in a given vicinity. Similarly roadway workers and train crews could broadcast information to multiple individuals simultaneously so as to coordinate activities and foster shared situation awareness. This selective broadcast capability would have the effect of reproducing the "common ground" that is fostered by the party-line feature of radio communication without contributing to the communication congestion associated with analogue radio where all messages are broadcast simultaneously to everyone.

*Summary*

This case study illustrates how informal collaborative practices evolve to increase system resilience. Railroad workers recognized a potential path to system failure, specifically the potential for collisions due to lack of mutual situation awareness. They have developed a variety of informal information extraction and proactive communication strategies to guard against this possibility. In designing new technologies it is important to understand the contribution of informal practices to system resilience so as to avoid the possibility of degrading system resilience by inadvertently disrupting effective informal work practices.

Effective support for distributed work require mechanisms to enable distributed parties to maintain awareness of the activities and plans of others so as to be able to coordinate goals, synchronize activities, prevent coordination breakdowns, and create resilience in the face of unanticipated events and errors (Hollnagel, Woods, and Leveson, 2006). New technologies such as position location and digital communication technology have the potential to facilitate communication and coordination across distributed organizations. However, if not carefully designed, they may disrupt existing strategies for building and maintaining the common ground that is critical to coordinating work and ensuring safe operations.

This study clarified how individuals in the distributed system worked to develop and maintain shared situation awareness and how shared situation awareness and informal cooperative strategies combined to facilitate work and enhance safety. It pointed to features that need to be incorporated in future systems to foster shared situation awareness and increase resilience of distributed work.

### Evolving Work Practices to Maintain Work System Resilience

The first case study illustrated how domain practitioners develop informal work practices that contribute substantively to system resilience. The second case study, drawn from a military command and control domain, reinforces and extends this finding by showing that practitioners continuously adapt their work practices in order to maintain resilience in the face of constantly changing world demands.

The second case study is based on field observations that were conducted in a command and control centre for a military airlift service organization. Periodic field observations and interviews were conducted over a seven-year period. We describe the changes in work demands that were observed over that period and the informal artefacts that users created to compensate for the growing mismatch that arose between their formal systems and the demands of the work.

*Background*

The airlift service organization plans, schedules, and tracks airlift missions to move military equipment and personnel worldwide. It is a global command centre with several hundred people planning and executing more than 350 missions per day. This includes a command and control centre that is responsible for the monitoring and control of flight missions starting 24 hours prior to planned mission launch, through to completion of the mission. The command and control centre is manned by command and control officers, enlisted controllers, and flight managers who are responsible for identifying, tracking, and resolving problems. They in turn are supported by a number of additional organizational groups providing specialized expertise including a group responsible for weather forecasting and tracking.

The research activity began in 2001 and has continued through to the present (2008). It has involved the development of a number of work-centred support systems (WCSS) intended to support the problem-solving and decision-making activities of personnel involved in scheduling and managing airlift resources. WCSS are designed to provide comprehensive support for the multiple aspects of work (e.g., decision support, product development support, collaborative support, and work management support) within an integrated work-oriented framework (Eggleston, 2003). The first system, called Work-Centered Support System for Global Weather Management (WCSS-GWM), was developed to support weather forecasting and monitoring in support of mission planning and execution. It is currently installed and in use in the airlift service organizations' command and control centre (Scott et al., 2005). More recently analysis and design activities have expanded to cover the larger airlift mission planning and execution process (Roth et al., 2007; Roth, Stilson et al., 2006).

Periodic field observations and structured interviews were conducted in the command and control centre as part of the WCSS development process. These began prior to development of the initial WCSS-GWM design concept so as to ground the design in the field of practice. Additional field observations occurred after the WCSS-GWM system was first deployed (toward the end of the second year), so as to uncover and address any unanticipated elements of work. Field observations and interviews have continued and expanded as part of additional WCSS prototype development efforts to support different positions within the command and control centre. The research has encompassed virtually all the staff positions in the command and control centre, including weather forecasters, flight managers, enlisted controllers, and command and control officers.

A variety of changes occurred over the seven-year period of the study that impacted on cognitive and collaborative work demands in the command and control centre. One of the most striking changes was in scale of operation. As work on the program was starting, in February 2001, the position of flight manager was just being created and staffed. The flight managers were only assigned a small percentage of the flights handled by the command and control centre. Initially, there was an average of three flight managers per shift and flight managers handled less than 20 flights a month. By February 2004 there was an average of 10 flight managers per shift and flight managers handled more than 3,000 flights a month. These numbers have continued to rise.

In response domain practitioners adapted their work practices and created informal artefacts to compensate for the growing mismatch that arose between their formal systems and the demands of the work. Below we summarize the types of changes that arose in work demands, and changes in work practice that were observed to take place in response. As in the first case study, these informal work practices contributed to the smooth functioning and resilience of the work system. A more complete description of the study results is provided in Roth, Scott, Deutsch et al., 2006.

*Evolving User Practices for Coping with a Changing World*

Over the seven-year period, changes were seen in:

- goals and priorities of the work (e.g., the nature of flight missions that were conducted; the parts of the world where missions operated);
- scale of operations;
- roles, team and organizational structure (e.g., new positions were created and there were shifts in the distribution of work across positions);
- complexity of problems faced (as the number of missions increased the airlift service organization hit against hard resource constraints making it more important to anticipate and respond to resource bottlenecks and prioritize among missions in cases of goal conflict);
- information sources and computer-support systems provided to support work;
- physical layout of the operations centre (the operations centre was remodelled with the result that weather forecasters were no longer in as close physical proximity with the flight managers they supported).

While some of the changes were anticipated, others were not. For example, the emergence of conflicts in Iraq and Afghanistan dramatically increased demands on available resources – both manpower and aircraft. Further, even in the case of anticipated changes, their impact on team roles and work structure were not necessarily foreseeable.

At the time that the WCSS-GWM was being developed, the organization anticipated, and informed us, that there would be a dramatic increase in the number of missions that would need to be handled, and a corresponding increase in staffing. However, while they anticipated an increase in scale, the management of the organization had not determined what changes would be needed in organizational structure to accommodate the increased number of missions. The new organizational structure that was eventually adopted could not have been foreseen ahead of time, as it emerged gradually, in an attempt to distribute work across positions so as to maximize productivity, while maintaining safety margins.

With the increase in scale there turned out to be a shift in team member roles and tasks. For example, while initially a weather forecaster worked one-on-one with a flight manager to produce a tailored forecast for each flight managed mission, the nature of the collaboration between forecaster and flight manager changed as the number of flight managers and flight-managed missions increased. There became three separate forecaster positions, one position generating forecasts for different geographic regions, one monitoring "high risk" missions, and one responsible for monitoring the remaining lower-risk missions.

Among the consequences of the various operational changes that we observed was a growing mismatch between the support provided by the formal information systems in place and the requirements of work. User requests for software modifications, however simple, required lengthy lead times on the order of months to years to satisfy. As a consequence, we observed users turn to the development of informal artefacts, including "home-grown" software, to compensate for system–work mismatches.

Over the course of our field observations we identified a number of cases where informal artefacts were created to compensate for the limitations and rigidity of existing information systems. These took the form of:

- physical artefacts such as handwritten cheat sheets and sticky notes;
- new visualizations that graphically depicted important information that was not provided by the information systems as designed;
- "local" databases that stored updates and corrections to information stored in the formal system databases;
- new software tools programmed by members of the user community to create support systems for aspects of work that were not well supported by the formal information systems.

The most striking cases of informal artefacts were instances where the user community developed their own "home-grown" software tools. These were developed to support aspects of work that were not well supported by the formal software systems developed and maintained by the larger organization. We observed three clear examples of these "home-grown" software tools, one created by the weather forecasting staff and two created by the command and control staff. In all three cases the tools were built by a member of the user community using "off-the-shelf" spreadsheet, drawing and word processing software. Macros were used to import data from the operational information systems, process and integrate it with locally available information, and then create new displays that better supported the work processes.

One example was a tool developed by weather forecasters to help them identify and track monitoring priorities across missions. The large increase in the number of missions handled meant that weather forecasters could no longer closely watch all missions individually. As a consequence, they needed to be able to identify "high risk" missions that would need closer attention. They devised a means to classify missions into different risk-level categories based on a combination of weather-related criteria. There were no provisions in the existing information systems for defining, displaying, or using these risk levels. Consequently, one of the forecasters developed a spreadsheet program to classify and manage missions by risk level. This software was used to allocate missions across weather forecasters, and focus attention on the "higher risk" missions that needed to be more closely monitored.

Command and control staff also created their own software tools to support situation awareness, help focus their attention on missions that need to be more closely watched, and manage their workload. In one case they developed a tool to help them closely track missions that were considered to be "high visibility" or that had problems (e.g., missions delayed due to maintenance problems). They created a "notepad" tool using standard word processing software with macros that allowed them to import information about these missions from the formal information systems, and add detailed annotations as to the current status of the missions and planned actions. Macros allowed the notepads to be periodically updated so that the user could be alerted to new problems. These notepads served as a focused "to do" list for the user, enabling them to prioritize and manage their work. It also served as a shift turnover log, allowing critical information to be shared across shifts, facilitating across-shift coordination and collaboration.

In a second case, the command and control staff needed a "big picture" view of the aircraft and missions under their control. They needed a means to rapidly assess what missions were assigned to which aircraft, and whether there were opportunities to add missions or shift missions across aircraft in order to increase throughput. The formal information systems provided tabular listings of missions that were ineffective in supporting these cognitive tasks. One of the command and control staff developed a graphic visualization to fill this gap in cognitive support.

The graphic display was created using a static drawing package (Visio). It displays missions assigned to aircraft on a timeline so that the user can easily see all aircraft under his or her control and what missions are assigned to each of those aircraft. The command and control staff uses drawing and "cut and paste" features to update the static display. In this way they have "cobbled" for themselves a "big picture" overview that allows them to see present and upcoming missions, their status and priorities all in a single graphic view. This allows them to readily answer incoming calls requesting information on mission status, to keep track of mission changes and communicate them across shifts, and to identify opportunities to add missions and re-task missions across aircraft so as to increase throughput.

These "home-grown" software artefacts provide salient examples of the creative work-arounds that users employ to compensate for mismatches between rigid software tools and the evolving demands of work. The informal tools and practices that emerged contributed substantively to the efficiency and resilience of the work system by enabling domain practitioners to maintain broad situation awareness, manage workload, and direct attention to high priority issues. These tools and practices were critical to the ability of the work system to cope with the rapid increases in workload and expanding demands on available resources.

The emergence of locally developed software artefacts such as new visualizations, local databases, and "home-grown" software tools is particularly noteworthy as these types of user-developed software "artefacts" have not been widely documented in the prior literature. They provide salient examples of the creative, and increasingly sophisticated, work-arounds that users employ to compensate for mismatches between rigid software tools and the evolving demands of work.

*Implications for Evolving Work-Centred Support Requirements*

The changes that arose in the command and control environment over the seven-year period of the study, and the resulting mismatches between the requirements of work and the software systems in place, make salient the need to develop software systems that can be more readily adapted to the changing requirements of work. This is critical to enable work systems to remain resilient in the face of a constantly changing world. We have coined the term "evolvable work-centred support systems" to describe the types of systems we envision (Roth, Scott, Deutsch et al., 2006). These evolvable work-centred systems would empower domain practitioners to modify and extend their software tools so as to more effectively support their work, enabling them to enhance productivity and margin for safety. Some technical ways forward for achieving this vision are documented in Roth, Scott, Deutsch et al. (2006).

*Summary*

This case study provides a compelling illustration that work practices do not remain fixed, but rather are constantly evolving in response to the changing demands of the world. Work support systems need to evolve correspondingly so as to maintain or enhance the level of resilience of the work system.

In this case study, there was a gradually increasing mismatch between the demands of the work and the formal information systems that were in place. Domain practitioners stepped in to bridge the gap. They developed their own software systems to support situation awareness, help direct attention, and manage workload. This enabled the organization to maintain resilience in the face of increasing workload and resource demand. The case study points to the need to incorporate "evolvable" features in software support systems to enable domain practitioners to more easily extend and adapt them to meet the changing requirements of work.

## Conclusions

The two case studies summarized above highlight the importance of informal work practices to overall system resilience. The studies illustrated situations where domain practitioners evolved informal work practices to enhance overall safety; as well as situations where domain practitioners developed "home-grown" tools to fill gaps in formal support systems, and maintain resilience in the face of increasing work demands.

The studies point to two important contributors to work system resilience. To achieve and maintain work system resilience, domain practitioners must be able to: (1) actively anticipate potential failure paths and adapt practices so as to avoid them; and (2) continuously evolve work practices so as to maintain operation within safe bounds while meeting efficiency/ productivity objectives.

System designers need to take these precepts into account when attempting to develop effective, work-centred support systems. To foster resilience, support systems need to accommodate both formal and informal work practices and to evolve to keep pace with changes in work practices that will inevitably arise in response to changing work demands.

A number of researchers have noted that users will informally tailor the design of their systems and work practices to better meet the local demands of the situation. This has been referred to as "finishing the design" (Mumaw et al., 2000; Vicente, 1999). The case studies presented here extend these ideas by emphasizing that the demands of the world are not fixed but will change over time. The resilience of a work system depends on the ability of domain practitioners to adapt in the face of change (Nathanael and Marmaras, 2008). Thus, "finishing the design" not only entails tailoring tools and practice to local conditions, it also entails extending and modifying work systems over time. The challenge facing the design community is how to empower domain practitioners to more readily evolve their tools and practices and share them with others so as to adapt to the changing demands of the world.

# Chapter 13

# The Infusion Device as a Source of Healthcare Resilience[1]

Christopher P. Nemeth
Michael O'Connor
Richard I. Cook

## Introduction

As a service sector, healthcare relies on the timely use of accurate information. While information technology (IT) has been advocated as a means to improve healthcare efficiency, safety, and reliability at the sharp (operator) end, recent reports of unexpected results with information systems (e.g., Xia and Lee, 2004; Ash et al., 2004) indicate that IT is often poorly suited to support healthcare cognitive work. These unexpected outcomes due to automation surprises indicate that IT demonstrates *brittle* properties (Sarter et al., 1997) that result from poor understanding of the work settings that they are intended to support. If IT systems are intended to support clinical work, they must match the same complexity as the work domain that they are intended to aid (Ashby, 1956). A better approach is called for to understand the nature of systems and their ability to perform and survive under duress: in other words, to be *resilient*.

One goal of research, design, and development is to link the technological capability with the adaptive power of people as goal-directed agents (Alexander, 1977). People actively manage the dynamic characteristics of their workplace by drawing on a deep knowledge of their work domain to create and use artefacts

1    Dr Nemeth's research has been supported by the Agency for Healthcare Research and Quality, and the US Food and Drug Administration (USFDA) Center for Device and Radiological Health (CDRH). The authors are grateful to David Mendonça for his insightful comments on draft versions of this chapter.

(Blumer, 1986). Workers create cognitive artefacts (Hutchins, 2002) in physical form (e.g., order forms, checklists, schedules) and digital form (e.g., equipment control and display interfaces, information) to aid their cognitive work. Prior work has shown how these artefacts can be used to understand (Xiao et al., 2001) and derive design guidance for IT systems to support such work settings, because the artefacts embody only the essential elements of a work domain (Nemeth, 2003; Nemeth et al., 2006). This makes it possible to pursue a design approach *from* the user *to* the system.

Butler and Gray (2006) have suggested IT as a means to improve mindfulness in the face of complex technologies and surprising environments. Assuming that is true, how can IT systems be configured in order to support such an approach? Klein, et al. (2004) propose traits that IT systems need in order to support joint activity: extended actions that are carried out by an ensemble of people who are coordinating with each other. Their ten challenges for automation to participate in joint activity (Table 13.1) set a longer-term agenda for IT system development. Six of the challenges, pertain to the infusion device example in this chapter:

1. Fulfil the requirements of a Basic Compact to engage in "common grounding" activities – an agreement to facilitate coordination, to work toward shared goals, and to prevent team coordination breakdowns.
2. Adequately model other participants' actions *vis-à-vis* the joint activity's state and evolution – the ability to coherently manage mutual responsibilities and commitments, in order to facilitate recovery from unanticipated problems.
3. Be mutually predictable – support the mental act of seeing ahead, with the frequent practical implication of preparing for what will happen.
4. Be directable – make it possible to deliberately assess and modify others' actions as conditions and priorities change.
5. Make pertinent aspects of their status and intentions obvious to their teammates – make targets, states, capacities, intentions, changes, and upcoming actions obvious.
6. Enable a collaborative approach – design each element of an "autonomous" system to facilitate the kind of give-and-take that characterizes natural and effective teamwork among groups of people.

## Table 13.1   Challenges for automation
*Source: From Klein et al., 2004.*

| | Challenge | Description |
|---|---|---|
| 1 | Fulfil the requirements of a Basic Compact | Engage in "common grounding" activities – an agreement to facilitate coordination, to work toward shared goals, and to prevent team coordination breakdowns |
| 2 | Adequately model other participants' actions vis-à-vis the joint activity's state and evolution | Coherently manage mutual responsibilities and commitments to facilitate recovery from unanticipated problems |
| 3 | Be mutually predictable | The mental act of seeing ahead, with the frequent practical implication of preparing for what will happen |
| 4 | Be directable | Able to deliberately assess and modify others' actions as conditions and priorities change |
| 5 | Make pertinent aspects of their status and intentions obvious to their teammates | Make targets, states, capacities, intentions, changes, and upcoming actions obvious |
| 6 | Observe and interpret signals of status and intentions | Agents that receive signals must be able to interpret the signals and form models of their teammates |
| 7 | Engage in negotiation | Intelligent agents must convey their current and potential goals so that appropriate team members can participate in the negotiations |
| 8 | Enable a collaborative approach | Every element of an "autonomous" system will have to be designed to facilitate the kind of give-and-take that quintessentially characterizes natural and effective teamwork among groups of people |
| 9 | Participate in managing attention | Team members direct each other's attention to the most important signals, activities, and changes, and must do this in an intelligent and context-sensitive manner, so as not to overwhelm others with low-level messages containing minimal signals mixed with a great deal of distracting noise |
| 10 | Help to control the costs of coordinated activity | Partners in a coordination transaction must do what they reasonably can to keep coordination costs down, to try to achieve economy of effort |

What is the value of automation as a "team player"? Improved ability to collaborate implies improved ability to respond to changes in the work setting – to *adapt*. Adaptability is at the heart of resilience, "the intrinsic ability of a system to adjust its functioning prior to, during, or following changes and disturbances" (Erik Hollnagel, personal communication, 24 June 2008). Patients in fragile health can change rapidly. This requires equipment, procedures, and staff that can adapt diagnostic and therapeutic interventions to best meet patient needs. The following example is based on our five-year study of infusion devices (Nunnally et al., 2004). It shows how resilience engineering (Hollnagel, Woods, and Leveson, 2006) can support adaptive decision-making by aiding the cognitive work of programming infusions.

## Clinical Cognitive Work

Healthcare relies on the accurate, timely description of process and condition. The cognitive work that clinicians undertake is necessarily contingent and interdependent. Clinical cognitive work is distributed among physicians, nurses, and pharmacists who are responsible for a patient's care, and is performed with information that is nearly always incomplete. Multiple individuals care for multiple patients, work without synchronizing with others, trade-off among priorities, and accommodate interruptions.

At the level of the individual, acute care patient, multiple diagnostic and therapeutic processes happen at the same time. Each patient's condition can be described by a spectrum of variables that are interrelated, and the interactions among those variables exceed the ability of clinicians to perceive them. Despite uncertain information, the clinician *must* act on behalf of the clinically ill patient. This compels the clinician to pursue diagnostic and therapeutic interventions that may be performed concurrently rather than sequentially. The decision to proceed with a certain treatment relies in part on the trade-offs between what is known about certain courses of treatment and their anticipated risks and benefits. The majority of these activities do not occur in what could be described as familiar territory, in which the data are sufficient and the patient's recovery is certain. Instead, patient condition and prognosis often exist in the kind of circumstances in which the available evidence on what to do is weak (Sharpe and Faden, 1998). Some practitioners contend that much of medical practice takes place where there is little proven knowledge and anticipated harms/benefits are unknown (Nemeth, 2005).

At the unit level (among and across patients), system performance in healthcare relies on the ability to match the demand for care with the resources to provide it. Care demand varies widely in amount, timing, and type. Patient condition and diagnoses, as well as their treatment, are highly specific to each individual. Resources such as sophisticated equipment and clinicians are limited by practical considerations such as qualifications and cost. This creates a work domain that relies on highly constrained resources to serve widely varying uncertain demand for the services it provides. Even under the best circumstances, an irreducible uncertainty makes it hard for clinicians to fully grasp the phenomena for which they are accountable. Our understanding of the biological system upon which clinicians intervene is woefully incomplete. At the unit level, recent increases in coordination demands due to staff resource limits (ACGME, 2002) place even greater need for the reliable exchange of information in instances such as between-shift hand-offs.

Success in the healthcare work setting depends on adaptability in the face of change. Schulman (1993b) contends having sufficient rational resources (*conceptual slack*) available makes it possible to cope with work domain constraints. Acute care clinicians have created their own rational resources, such as advance planning calendars and call schedules, to manage such workplace complexity (Nemeth et al., 2007). The design of equipment, though, is largely in the hands of manufacturers.

## Support for Clinical Cognitive Work

In any organization, workers and managers need information on changing vulnerabilities and new resources to manage them. IT's inherent flexibility can be exploited to support this cognitive work by supporting information access, retrieval, display, communication, and cognitive aiding.

IT systems were initially used at the blunt (management) end of healthcare organizations to support billing and patient records. This made it possible in the 1990s to shift workload to portions of the organization that had available productive capacity. The move had the simultaneous effect of also shifting risks, as well as consequences, elsewhere in the organization (Sarter et al., 1997). Such shifting produced new interactions among individuals and groups. The number of possible interactions increased and remained unknowable, and no real tools were created to manage the results of these new interactions. Indeed, IT systems are often installed in an attempt to fix problems that are actually embedded in the social organization (Wears and Berg, 2005). As long as the scale and stakes were small, the unanticipated results from such subtle changes could be tolerated as inconvenient interruptions in the clerical and management setting. However, the cognitive work and stakes at the sharp end are far different than at its blunt end (Nemeth et al., 2005). Misperceptions about user-device interaction have substantial consequences for clinical work.

In healthcare, the presentation of information directly influences clinicians' ability to develop an effective mental representation of past, current, and prospective states of patients. Representations of data that simultaneously show both constraints and possible solutions for dealing with those constraints are crucial to orient and re-orient clinicians. Systems that fit well with the work that is being done are highly useful. Present IT from equipment to information systems do not fit clinical work, though, and fall short in support of it.

Recent developments have made ever greater amounts of data related to patients available to clinicians. Just because data are available does not mean they are useful. In order to be useful, data must be manageable. This simple statement belies the depth and complexity that is involved in the clinical work setting. Care providers exist in an information ecology that includes the patient, other clinicians, devices, information systems, and physical artefacts. Figure 13.1 shows the current state of support for clinical healthcare cognitive work. Care providers attend to individual patients (shown as direct contact by black lines) using their own observation, consultant views, and patient self-reports. They direct and monitor therapeutic and diagnostic interactions (through indirect contact shown by dotted lines) that is shown in the lower portion of Figure 13.1. Medical devices are increasingly connected to communication networks to "push" significant information to clinicians and related systems. Clinicians consult physical cognitive artefacts (Hutchins, 2000, 2002) such as paper charts, orders, and status boards. They request and synthesize data from a variety of information systems and departments (shown in the upper portion of Figure 13.1). All this happens in the context of caring for multiple patients who each have unique needs and care trajectories that must be planned, coordinated, executed, evaluated, and assessed. Three elements in this ecology give us an example of the current state of support for care provider cognition: medical records, decision aids, and medical devices.

Electronic versions of *medical records* (EMR) attempt to make the large amount of information that they contain useable. Despite these efforts, clinicians find the EMR is a poor match for the kinds of cognitive work they perform. This mismatch arises from increasing regulatory requirement that the medical record support billing for clinical activity. This configuration of records to support billing and not clinical purposes causes difficulty in locating critical information within the vast amount of information that the record contains, and the inability to use the record for important clinical activities such as the comparison of data. Now that it no longer serves a clinical role, clinicians have resorted to performing additional work to create their own informal and parallel system: the sign-out sheet. Each shift, clinicians list each of the patients on a unit along with critical items of information that are related to their condition and care (Nemeth et al., 2006).

Clinical *decision aids* (shown in the upper portion of Figure 13.1) have sought to help physicians synthesize complex considerations into rule-based guidance about patient care decisions. Berg (1997) describes how such computer-based approaches that are intended to support clinician cognitive work produce mixed results. Decision support systems need to be constantly monitored to determine whether their suggestions conform with practice. In some instances, the number of branching points that are needed to accommodate exceptions may become so great that the system is impossible to use and maintain (Ash et al., 2004). The failure of this approach demonstrates that decision-making under

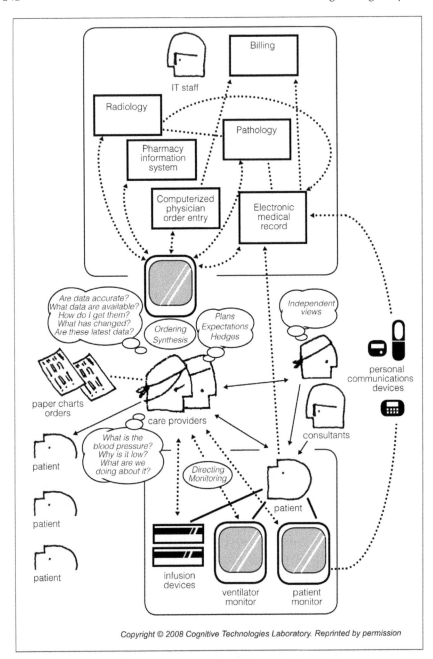

Copyright © 2008 Cognitive Technologies Laboratory. Reprinted by permission

**Figure 13.1   Clinician information ecology**

*Copyright © 2008 Cognitive Technologies Laboratory. All rights reserved.*

clinical conditions is far more complex and less tractable than proponents of these early systems believed. As a result, clinical decision support systems' effect on practitioner performance and patient health remain as inconsistent as they were 15 years ago (Garg et al., 2005). Relatively few clinical decision support systems (CDSS) are in use after their introduction over 25 years ago (Kaplan, 2001).

Berg's research also yielded a critical observation about IT design. In healthcare, systems that fit poorly with work rapidly become useless when circumstances change through either the evolution of care or degeneration of the system. As a result, IT must conform to the work that is being done in order to support resilience.

Medical *devices* such as infusion pumps increasingly feature complex control and display interfaces. A large number of pumps, as many as nine to twenty, can be necessary to support a single patient in the intensive care unit (ICU) or emergency department (ED). Even highly experienced clinicians who have used infusion devices for years get "lost in menuspace" when they perform even the simplest tasks (Nunnally et al., 2004). Figure 13.2 depicts the questions in the mind of a clinician that an infusion device display needs to answer, but does not. This is because contemporary infusion device displays are limited to showing only current status. They neither account for events that preceded the current state, nor indicate what to expect or what options might be possible or advisable in the future. Appreciating both the context from past occurrences and potential for future events are essential to making informed patient care decisions. IT that closely supports this cognitive work would support healthcare resilience.

## Developing Healthcare IT

"Service oriented architectures" (SOA) are the software development community's current approach to providing user-oriented IT. SOA was conceived as the way to provide web-based services that are tailored to the needs of service users. This is intended to be done through flexible, standardized modules to connect applications and data (Papazoglou, 2003). SOA can be a

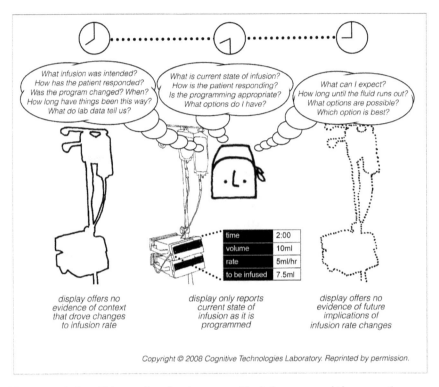

**Figure 13.2    Disparity between clinician cognition and pump display**

way of thinking about building software, both as an architecture model and as a programming model (Channabasavaiah et al., 2004). The information ecology in Figure 13.1 suggests that SOA could play a role in healthcare IT support by bridging the wide variety of platforms that carry clinical information. At the moment, though, SOAs are developed *from* the system *to* the user (see Agrawal et al., 2002; Haller et al., 2005). In order to optimize the use of software modules, SOAs rely on a static user context. This is difficult to reconcile with any work application that is fluid (Perrey and Lycett, 2003). The problem becomes even more pronounced when attempting to support complex, dynamic work settings that are poorly bounded and highly contingent, such as healthcare.

If SOAs work, we should get well-tailored systems that serve users – but we do not. Why is that? It is not due to a lack of interest, but rather a lack of insight. Experience shows that user-guided IT development is a failure. This is because expert operators are poor objective sources of information on their own work, and lack the skills to study what is necessary to develop well-articulated requirements for IT support. For example, clinical practitioners are poor judges of their own diagnostic reasoning (McNutt et al., 2005) and have difficulty with objective descriptions and analysis of this crucial cognitive task. Instead, insight into actual work processes that are performed in the real world comes from researchers who study human performance (see Nemeth et al., 2004; Nemeth, 2007).

The following example demonstrates how IT can follow these principles through the design of an infusion pump control-display interface.

## An Example of Healthcare IT to Improve Resilience

Most infusions in US hospitals are now provided by infusion devices (Hunt-Smith et al., 1999), making it the most widely used IT-controlled equipment in the acute care environment. While it is convenient to think in terms of *an* infusion device, the nature of infusion requires a complex collection of elements that depend on each other to work together reliably. Figure 13.3 depicts these elements as a socio-technical system in the same hierarchy that Rasmussen (1997a) used to describe system safety. Government, regulators, and associations develop policy and guidance that are intended to coordinate the activities of company-level organizations such as manufacturers and healthcare facilities. Management and staff collaborate across departmental, temporal, and geographic barriers to deliver infusions. It is at the level of work, though, where patient care and safety play out minute-by-minute that the design of equipment can have an effect. Problems with commercially available infusion devices arise from mismatch between the complexity of clinical care and the simplicity of infusion devices' limited "keyhole" display and keypad. Most cell phones have more keys than infusion devices. The results of the mismatch can be severe, as microprocessor-based infusion devices are associated with significant clinical accidents that result in patient morbidity and mortality (Nullally et al., 2004). These adverse outcomes demonstrate the brittle nature of IT support for infusions, and the need for a solution that better fits clinical work.

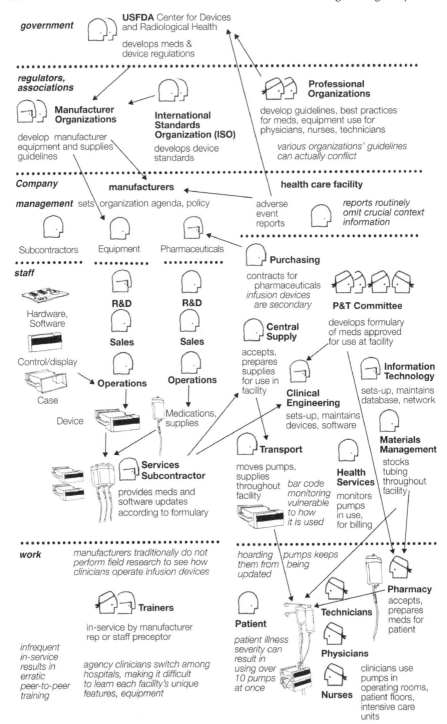

**government**  USFDA Center for Devices and Radiological Health

develops meds & device regulations

**regulators, associations**

Manufacturer Organizations

develop manufacturer equipment and supplies guidelines

International Standards Organization (ISO)

develops device standards

Professional Organizations

develop guidelines, best practices for meds, equipment use for physicians, nurses, technicians

*various organizations' guidelines can actually conflict*

**Company**  manufacturers                    health care facility

**management** sets organization agenda, policy

adverse event reports

reports routinely omit crucial context information

Subcontractors  Equipment  Pharmaceuticals

Purchasing

contracts for pharmaceuticals *infusion devices are secondary*

P&T Committee

**staff**

Hardware, Software

R&D           R&D

Sales         Sales

Control/display

Operations    Operations

Case

Central Supply

develops formulary of meds approved for use at facility

Information Technology

sets-up, maintains database, network

Device

accepts, prepares supplies for use in facility

Clinical Engineering

sets-up, maintains devices, software

Materials Management

Medications, supplies

Services Subcontractor

provides meds and software updates according to formulary

Transport

moves pumps, supplies throughout facility

*bar code monitoring vulnerable to how it is used*

Health Services

monitors pumps in use, for billing

stocks tubing throughout facility

**work**  *manufacturers traditionally do not perform field research to see how clinicians operate infusion devices*

*hoarding pumps keeps them from being updated*

Trainers

in-service by manufacturer rep or staff preceptor

*infrequent in-service results in erratic peer-to-peer training*

*agency clinicians switch among hospitals, making it difficult to learn each facility's unique features, equipment*

**Patient**

*patient illness severity can result in using over 10 pumps at once*

Technicians

Physicians

Nurses

Pharmacy

accepts, prepares meds for patient

clinicians use pumps in operating rooms, patient floors, intensive care units

## Figure 13.3    Infusion device as a socio-technical system

Improving the compatibility between infusion pumps and work requirements is not a matter of fixing a particular aspect of a particular design, such as making display type larger. Instead, it requires a new approach to representation that aids the work of clinicians who perform infusions. A new design needs to follow Klein et al.'s principles in order to contribute to the resilience of the healthcare work setting. Making the pump's operation straightforward would make it directable. Making its programming state understandable would make status and intentions obvious, facilitate coordination, and provide a way for the operators to recover from unanticipated problems. Demonstrating implications of current programming for the future would make it predictable, enable the operator to foresee potential future states, and make upcoming actions obvious. Enabling others (in addition to the clinician who programmed the pump) to make informed decisions in light of this information supports collaboration and common grounding activities.

Figure 13.4, which reflects our lab's five-year study of commercially-available infusion devices, illustrates how an interface could provide information about device display and control through time. The design is organized into fields that

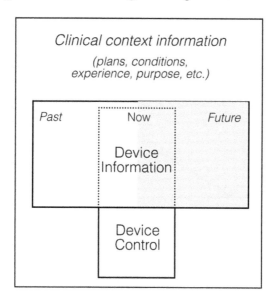

**Figure 13.4    Infusion device interface schematic**

contain infusion information (the diagram at centre), infusion control (at bottom), and context information (around the information and control fields). The display is organized to show volume/time (rate) parameters, current and past system status, and the expected course if current parameters are maintained. While the control position remains fixed, the infusion information "scrolls" from right to left of the diagram as time passes.

Flattening the interface hierarchy that current devices use implies a different approach to design. Current infusion devices present a few generic buttons that are used to navigate a deep hierarchy. This concept provides more controls with unique functions to traverse a shallower hierarchy, lessening the likelihood of mode error and disorientation.

In the following hypothetical example, a 4.6 kg infant is receiving a planned volumetric infusion of 10 per cent dextrose that was started at 08:02 and is programmed to be completed in two hours, at 10:02. Figure 13.5 shows how the interface would

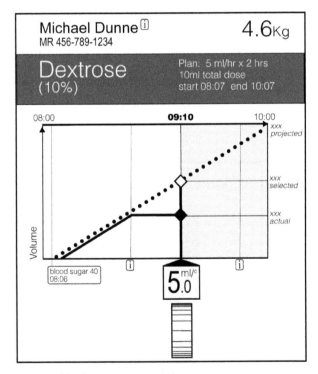

**Figure 13.5    Infusion device interface supporting resilience**

look 30 minutes into the infusion. Operating history, current state, and implications for the future appear within the same display. Additional information (indicated by "i" symbols) can be opened and read. For example, the item on the lower left provides blood glucose test results that were reported at 08:06. Both items at the lower left-hand edge of the diagram could be opened to obtain more information on programming that was done at 08:00 and 08:45. The display shows volume/time (rate) parameters, current-past system status, and the expected course if current parameters are maintained. The "thumb wheel" virtual control at lower centre would make it possible for a clinician to adjust the rate of infusion. Moving the control up or down would adjust the rate (indicated by the dotted diagonal line) to various settings. Values for each variable such as the actual setting that are displayed at the edges of the diagram would change to indicate the implications of various rate changes. This would make it possible for a clinician to examine different rate settings and to choose which one best fits the patient's needs. It also shows what would happen in the future if a rate is chosen. After examining the various options and making a choice, the clinician can select it and the pump would then change the rate.

Certain clinical procedures can require that infusions be paused, then resumed. Figure 13.6 shows the data space for this infusion example as it was started, then paused for 15 minutes. Pausing the infusion for 15 minutes would result in a 1.5 ml deficit (indicated in the diagram). If the total volume to be infused is still to be completed within two hours, the rate will need to be adjusted to make up for the 1.5 ml deficit the pause caused. Figure 13.7 shows the evolution of the interface through this sequence.

The graphic representation makes it possible for clinicians to use pattern recognition to determine how infusions are programmed and progressing. Alphanumeric characters provide values for specific discrete variables that are necessary for accuracy. The display is predictive, demonstrating what will happen in the future if present parameters are maintained. This makes it possible for an operator to immediately recognize the kinds of dose limit errors that plague current infusion displays that are programmed using only numbers. The display would reflect both the infusion and context information changes through time. The

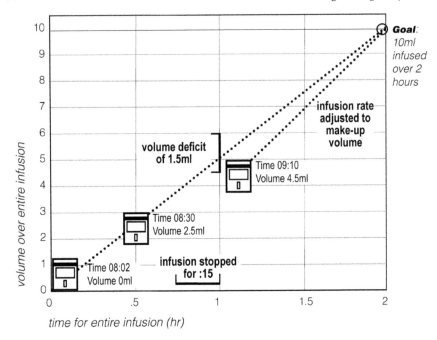

**Figure 13.6    Overview of hypothetical two-hour infusion**

*Copyright © 2008 Cognitive Technologies Laboratory. All rights reserved.*

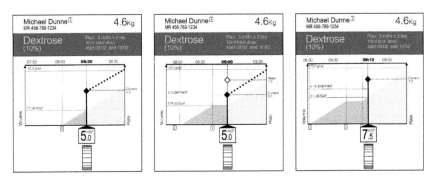

**Figure 13.7    Display at three points along two-hour infusion**

*Copyright © 2008 Cognitive Technologies Laboratory. All rights reserved.*

infusion information field shows fluid volume already delivered, and fluid volume to be delivered at the rate that is selected.

Figure 13.8 provides a further example of how such a display can present a summary of treatment and intentions that is understandable at a glance. Shock is a serious, life-threatening condition in which insufficient blood flow reaches the body tissues. Reduced blood flow hinders the delivery of oxygen and

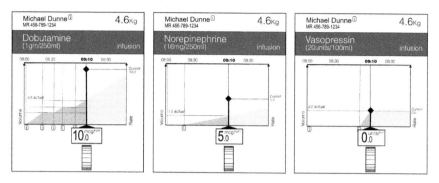

**Figure 13.8    Three infusion displays while treating a pediatric patient in shock**

nutrients to the tissues. Treatment for an acute care patient who is in shock will typically start with an infusion of Dobutamine to increase cardiac contractility and output. Norepinephrine, and later Vasopressin, can also be started to increase blood pressure. Figure 13.8 shows how the infusion display discussed here provides a recognizable pattern to any clinician who would arrive at a patient's bedside. At a glance, it is clear the patient was started on an infusion of Dobutamine at 8:00am that was then titrated (adjusted) upward. Norepinephrine was started at 8:20 and titrated to improve blood pressure. Vasopressin was begun at 9:00 to narrow blood vessel walls, augmenting the effect of the other two drugs. The displays demonstrate the value of being able to understand the recent past with a minimum of ambiguity. They also enable the clinician to make well-informed decisions about the future course of treatment.

Rather than a narrow "keyhole" display showing only current system state, the display in Figures 13.5, 13.7, and 13.8 provide context and indicate implications for the future. Including information on the clinical context makes it possible to interpret device behavior in terms of its clinical use.

## Discussion

In our concept, the state of the device reflects the state of the world in which the device is operating. This makes it possible for the clinician to make informed decisions regarding device performance in the context of patient care. These are the kinds of observable and controllable traits that would improve IT support for healthcare system resilience (Nemeth et al., 2008). The concept reflects many of Klein, et al.'s challenges, making it better suited to participate in joint activity with clinicians.

Making clinical and programming information explicit facilitates coordination and helps to prevent team coordination breakdowns. Providing past, current, and anticipated states, and making connections with related data such as lab results, facilitate recovery from unanticipated problems. Showing projected values supports the mental act of seeing ahead to assist preparation for what will happen. Controls that make it possible to explore contingencies before committing to a final decision enable the clinician to evaluate trade-off decisions. Integrating controls with displayed information makes it possible to deliberately assess and modify programmed actions as conditions and priorities change. The combination of graphic and alphanumeric information makes pertinent aspects of the device target, status, capacities, programming intentions, and upcoming actions obvious to members of the clinical team.

Improvements to infusion device interface design as we described above can make infusion at the work level more resilient in *preparation* and *restoration*. In terms of *preparation*, the proposed concept would enable clinicians to anticipate the needed flow rate by using the predictive display to extrapolate into the future. Including real-world data such as lab results along with the history of the device's programming provides context for decisions about future actions that might be taken. The overlay also ensures that perceptions are grounded in understanding what results are actually being obtained. Decisions about what to program rely on trials and speculation about what might occur if various courses of action were taken. Allowing the clinician to explore and compare alternative infusion programs without having to commit makes it possible to consider trade-offs well before deciding on a course of action. The clinician can see the implications of the change that is being planned before making it. The ability to anticipate possible complications before they occur enables the clinician to try out options without incurring a possible loss.

In terms of *restoration*, the display presents observable data through time. This makes it possible to re-establish what has happened after an infusion was stopped. A graphical display enables the operator to recognize what has happened not just to the pump, but to the patient. The ability to look forward and backward through time makes it possible to review the infusion state at crucial times, such as between-shift hand-offs, when it is essential to know how the patient and treatment are faring. Each of these features supports restoration by making it possible to look at what actually happened in context.

## Summary

While systems that are a poor fit with clinical work make it brittle, IT that fits clinical work contributes to improved resilience. Research into resilience is a promising avenue to influence the course of development for healthcare IT by addressing questions that have genuine import for healthcare and the systems that are intended to support it. Healthcare that is resilient readily adapts to changing demands. The essence of resilience is a system that conforms with the work that it is intended to support and remains useful as the work setting changes or degrades. IT systems have the potential to change rapidly and to convey needed information in the face of changes and challenges that clinicians face. The ability of those systems to realize that potential pivots on a valid understanding of the work domain that they are intended to support.

# Bibliography

AAIB (Air Accidents Investigation Branch) (1996), *Aircraft Incident Report 3/96: Report on the Incident to Boeing 737-400, G-OBMM near Daventry, on 23 February 1995* (London: HMSO).

ACGME (Accreditation Council for Graduate Medical Education) (2002), *Report of the ACGME Work Group on Resident Duty Hours* (Chicago: ACGME) 11 June.

Adamski, A.J., and Westrum, R. (2003), "Requisite Imagination: The Fine Art of Anticipating What Might Go Wrong," in E. Hollnagel (ed.), *Handbook of Cognitive Task Design* (Mahwah, NJ: Lawrence Erlbaum Associates).

Adger, W.N., Hughes, T.P., Folke, C., Carpenter, S.R., and Rockström, J. (2005), "Social-Ecological Resilience to Coastal Disasters," *Science* 309:5737, 1036–9.

Adler, P.S., and Kwon, S. (2000), "Social Capital: The Good, the Bad, and the Ugly," in E.L. Lesser (ed.), *Knowledge and Social Capital: Foundations and Applications* (Boston: Butterworth-Heinemann) 89–115.

Agrawal, R., Bayardo, R., Gruhl, D., and Papadimitriou, S. (2002), "Vinci: A Service Oriented Architecture for Rapid Development of Web Applications," *Computer Networks* 39:5, 523–9.

Aguirre, B.E. (2006), *On the Concept of Resilience* (Newark, DE: University of Delaware, Disaster Research Center).

Alesch, D.J., and Petak, W.J. (1986), *Politics and Economics of Earthquake Hazard Mitigation: Unreinforced Masonry Buildings in Southern California* (Boulder, CO: University of Colorado).

Alesch, D.J., and Petak, W.J. (2001), *Overcoming Obstacles to Implementation: Addressing Political, Institutional and Behavioral Problems in Earthquake Hazard Mitigation Policies* (Buffalo, NY: MCEER).

Alexander, C. (1977), *A Pattern Language: Buildings, Towns, and Cities* (New York: Oxford University Press).

Allen, J.T. (1997), "Defining Disaster Down," *Slate Magazine* (published online 18 January 1997) <http://www.slate.com/id/1883/>, accessed 22 March 2008.

Amabile, T.M. (1988), "A Model of Creativity and Innovation in Organizations," in B.M. Staw and L.L. Cummings (eds), *Research in Organizational Behavior, Vol. 10* (Greenwich, CT: JAI Press).

Amalberti, R. (2006), "Optimum System Safety and Optimum System Resilience: Agonistic or Antagonistic Concepts?" in E. Hollnagel, D.D. Woods and N.G. Leveson (eds), *Resilience Engineering: Concepts and Precepts* (Aldershot, UK: Ashgate) 253–71.

Anderies, J.M., Janssen, M.A., and Ostrom, E. (2004), "A Framework to Analyze the Robustness of Social-ecological Systems from an Institutional Perspective," *Ecology and Society* 9:1, 18 (published online 9 June 2004) <http://www.ecologyandsociety.org/vol9/iss1/art18>.

Anders, S., Woods D.D., Wears, R.L., Perry, S., and Patterson, E.S. (2006), "Limits on Adaptation: Modeling Resilience and Brittleness in Hospital Emergency Departments." Paper presented at the 2nd International Symposium on Resilience Engineering, Juan-les-Lins, France.

Andrews, R., Biggs, M., and Seidel, M. (eds) (1996), *The Columbia World of Quotations* (New York: Columbia University Press). Retrieved at Bartleby.com website <http://www.bartleby.com/66/4/40404.html>, accessed 20 October 2008.

Ash, J.S., Berg, M., and Coiera, E. (2004), "Some Unintended Consequences of Information Technology in Health Care: The Nature of Patient Care Information System-related Errors," *Journal of the American Medical Informatics Association* 11:2, 104–12.

Ashby, W.R. (1956), *An Introduction to Cybernetics* (London: Chapman & Hall).

Bainbridge, L. (1983), "Ironies of Automation," *Automatica* 19:6, 775–9.

Beer, S. (1981), *Brain of the Firm*, 2nd edn (New York: John Wiley & Sons).

Beer, S. (1985), *Diagnosing the System for Organizations* (New York: John Wiley & Sons).

Beinhocker, E.D. (1999), "Robust Adaptive Strategies," *Sloan Management Review* 40:3, 95–106.

Bengtsson, J., Angelstam, P., Elmqvist, T., Emanuelsson, U., Folke, C., Ihse, M., et al. (2003), "Reserves, Resilience and Dynamic Landscapes," *Ambio* 32:6, 389–96.

Benner, M.J., and Tushman, M.L. (2003), "Exploitation, Exploration, and Process Management: The Productivity Dilemma Revisited," *Academy of Management Review* 28: 238–56.

Berg, M. (1997), *Rationalizing Medical Work* (Cambridge, MA: The MIT Press).

Billings, C.E. (1996), *Aviation Automation: The Search for the Human-Centered Approach* (Hillsdale, NJ: Lawrence Erlbaum Associates).

Birkland, T.A. (2006), *Lessons of Disaster* (Washington, DC: Georgetown University Press).

Birkland, T.A., Burby, R.J., Conrad, D., Cortner, H., and Michener, W.K. (2003), "River Ecology and Flood Hazard Mitigation," *Natural Hazards Review* 4:1, 46–54.

Blumer, H. (1986), *Symbolic Interactionism: Perspective and Method* (Berkeley: University of California Press).

Bogner, M.S. (ed.) (1994), *Human Error in Medicine* (Hillsdale, NJ: Lawrence Erlbaum Associates).

Boisot, M., and Child, J. (1999), "Organizations as Adaptive Systems in Complex Environments: The Case of China," *Organization Science* 10:3, 237–52.

Bowser, H. (1987, Summer), "Maestros of Technology: An Interview with Arthur Squires," *American Heritage of Invention and Technology* 3:1, 24–30.

Boyd, T.A. (1957), *Professional Amateur: The Life of Charles Franklin Kettering* (New York: E.P. Dutton).

Branlat, M., Anders, S., Woods, D.D., and Patterson, E.S. (2008), "Detecting an Erroneous Plan: Does a System Allow for Effective Cross-checking?" in E. Hollnagel, C. Nemeth, and S. Dekker (eds), *Remaining Sensitive to the Possibility of Failure* (Aldershot, UK: Ashgate) 247–58.

Brinkley, D. (2006), *The Great Deluge: Hurricane Katrina, New Orleans, and the Mississippi Gulf Coast* (New York: HarperCollins).

Brown, S., and Eisenhardt, K.M. (1997), "The Art of Continuous Change: Linking Complexity Theory and Time-paced Evolution in Relentlessly Shifting Organizations," *Administrative Science Quarterly* 42:1, 1–34.

Burby, R.J. (1998a), *Cooperating with Nature: Confronting Natural Hazards with Land Use Planning for Sustainable Communities* (Washington, DC: Joseph Henry Press).

Burby, R.J. (1998b), "Natural Hazards and Land Use: An Introduction," in R.J. Burby (ed.), *Cooperating with Nature: Confronting Natural Hazards with Land-Use Planning for Sustainable Communities* (Washington, DC: Joseph Henry Press).

Burby, R.J. (2006), "Hurricane Katrina and the Paradoxes of Government Disaster Policy: Bringing about Wise Governmental Decisions for Hazardous Areas," *The Annals of the American Academy of Political and Social Science* 604:1, 171–91.

Burby, R.J., Beatley, T., Berke, P.R., Deyle, R.E., French, S.P., Godschalk, D., et al. (1999), "Unleashing the Power of Planning to Create Disaster-resistant Communities," *Journal of the American Planning Association* 65:3, 247–58.

Burby, R.J., and May, P.J. (1998), "Intergovernmental Environmental Planning: Addressing the Commitment Conundrum," *Journal of Environmental Planning and Management* 41:1, 95–111.

Burby, R.J, May, P.J., Berke, P., Dalton, L., French, S., and Kaiser, E. (1996), *Making Governments Plan: State Experiments in Managing Land Use* (Baltimore: Johns Hopkins University Press).

Burian, B.K., and Barshi, I. (2003), "Emergency and Abnormal Situations: A Review of ASRS Reports," *Proceedings of the 12th International*

*Symposium on Aviation Psychology* (Dayton, OH: Wright State University Press).

Butler, B.S., and Gray, P.H. (2006), "Reliability, Mindfulness, and Information Systems," *MIS Quarterly* 30:2, 211–24.

Cabbage, M., and Harwood, W. (2004), *Comm Check: The Final Flight of Shuttle Columbia* (New York: Simon and Schuster).

CAIB (*Columbia* Accident Investigation Board) (2003), Report, 6 vols. Government Printing Office, Washington, DC <http://caib.nasa.gov/news/report/default.html>.

Campanella, T.J. (2006), "Urban Resilience and the Recovery of New Orleans," *Journal of the American Planning Association* 72:2, 141–6.

Carlson, J.M., and Doyle, J. (1999), "Highly Optimized Tolerance: A Mechanism for Power Laws in Designed Systems," *Physical Review E* 60:2, 1412–27.

Carlson, J.M., and Doyle, J. (2002), "Complexity and Robustness," In *Proceedings of the National Academies of Science* 99:Supp 1, 2538–45.

Chaikin, A. (1995), *A Man on the Moon: The Triumphant Story of the Apollo Space Program* (New York: Penguin Group).

Channabasavaiah, K., Holley, K., Tuggle, E. (2004), "SOA is More than Web Services" <http://www.looselycoupled.com/opinion/2004/chann-soa-infr0630.html>.

Checkland, P. (1981), *Systems Thinking, Systems Practice* (Chichester, UK: John Wiley & Sons).

Checkland, P., and J. Scholes (1990), *Soft Systems Methodology in Action* (Chichester, UK: John Wiley & Sons).

Christensen, C.M. (1997), *The Innovator's Dilemma* (New York: HarperCollins).

Christoffersen, K., and Woods, D.D. (2002), "How to Make Automated Systems Team Players," *Advances in Human Performance and Cognitive Engineering Research* 2, 1–12.

Clarke, L. (2005a), *Worst Case Katrina*. Social Science Research Council 2005 (cited 22 November 2005) <http://understandingkatrina.ssrc.org/Clarke/>.

Clarke, L. (2005b), *Worst Cases: Terror and Catastrophe in the Popular Imagination* (Chicago: University of Chicago Press).

Coel, M. (1987), "A Bridge That Speaks for Itself," *American Heritage of Invention and Technology* 3, 8–17.

Collins, J.C., and Porras, J.I. (1994), *Built to Last: Successful Habits of Visionary Companies* (New York: Harper Business).

Comfort, L.K. (1994), "Risk and Resilience: Inter-Organizational Learning Following the Northridge Earthquake of 17 January 1994," *Journal of Contingencies and Crisis Management* 2:3, 157–70.

Comfort, L.K., and Haase, T.W. (2006), "Communication, Coherence and Collective Action: The Impact of Hurricane Katrina on Communications Infrastructure," *Public Works Management & Policy* 11:1, 6–16.

Conover, W.J. (1999), *Practical Nonparametric Statistics* (New York: John Wiley & Sons).

Cook, R.I. (2006), "Being Bumpable," in D.D. Woods and E. Hollnagel (eds), *Joint Cognitive Systems: Patterns in Cognitive Systems Engineering* (Boca Raton, FL: CRC Press, Francis & Taylor).

Cook, R.I., and Nemeth, C. (2006), "Taking Things in One's Stride: Cognitive Features of Two Resilient Performances," in E. Hollnagel, D.D. Woods and N. Leveson (eds), *Resilience Engineering: Concepts and Precepts* (Aldershot, UK: Ashgate).

Cook, R., and Rasmussen, J. (2005), "'Going Solid': A Model of System Dynamics and Consequences for Patient Safety," *Quality and Safety in Health Care* 14:2, 130–4.

Cooper, C., and Block, R. (2006), *Disaster: Hurricane Katrina and the Failure of Homeland Security* (New York: Times Books).

Coutu, D.L. (2002), "How Resilience Works," *Harvard Business Review* 80:5, 46–55.

Cropley, A. (2006), "In Praise of Convergent Thinking," *Creativity Research Journal* 18:3, 391–404.

Cross, R., Liedtka, J., and Weiss, L. (2005), "A Practical Guide to Social Networks," *Harvard Business Review* 83:3, 124–32.

Cross, R., and Prusak, L. (2002), "People Who Make Organizations Go – Or Stop," *Harvard Business Review* 80:6, 5–12.

Crossan, M.M., Lane, H.W., and White, R.E. (1999), "An Organizational Unlearning Framework: From Intuition to Institution," *Academy of Management Review* 23:3, 522–37.

Csete, M.E., and Doyle, J.C. (2002), "Reverse Engineering of Biological Complexity," *Science* 295:5560, 1664–9.

Cullen (2000), *The Ladbroke Grove Rail Inquiry* (Norwich, UK: HSE Books).

Cutter, S.L., and Emrich, C.T. (2006), "Moral Hazard, Social Catastrophe: The Changing Face of Vulnerability along the Hurricane Coasts," *The Annals of the American Academy of Political and Social Science* 604:1, 102–12.

D'Aveni, R.A. (1994), *Hypercompetition: Managing the Dynamics of Strategic Maneuvering* (New York: Free Press).

D'Aveni, R. (1999), "Strategic Supremacy through Disruption and Dominance," *Sloan Management Review* 40:3, 127–35.

Davis, M. (1998), *Ecology of Fear: Los Angeles and the Imagination of Disaster* (New York: Metropolitan Books).

Dawes, S.S., Birkland, T., Tayi, G.K., and Schneider, C.A. (2004), *Information, Technology and Coordination: Lessons from the World Trade Center Response* (Albany, NY: Center for Technology in Government, State University of New York).

Dean, C. (1996), "Is It Worth It to Rebuild a Beach? Panel's Answer Is a Tentative Yes," *New York Times*, 2 April, C4.

Deevy, E. (1995), *Creating the Resilient Organization: A Rapid Response Management Program* (Englewood Cliffs, NJ: Prentice Hall).

Dekker, S.W.A. (2003), "Failing to Adapt or Adaptations that Fail: Contrasting Models on Procedures and Safety," *Applied Ergonomics* 34:3, 233–8.

Dekker, S.W.A. (2006), "Resilience Engineering: Chronicling the Emergence of a Confused Consensus," in E. Hollnagel, D.D. Woods, and N. Leveson (eds), *Resilience Engineering: Concepts and Precepts* (Aldershot, UK: Ashgate).

Dekker, S.W.A., and Woods, D.D. (1999), "To Intervene or Not to Intervene: The Dilemma of Management by Exception," *Cognition, Technology & Work* 1:2, 86–96.

Dismukes, R.K., Berman, B.A., and Loukopoulos, L.D. (2007), *The Limits of Expertise: Rethinking Pilot Error and the Causes of Airline Accidents* (Aldershot, UK: Ashgate).

Doz, Y., and Kosonen, M. (2007), "Strategic Renewal: Building Strategic Agility," *International Strategic Management Society Conference* (San Diego, CA).

Dutton, J.E., and Jackson, S.E. (1987), "Categorizing Strategic Issues: Links to Organizational Action," *Academy of Management Review* 12:1, 76–90.

Dynes, R.R. (2003), "Finding Order in Disorder: Continuities in the 9/11 Response," *International Journal of Mass Emergencies and Disasters* 21:3, 9–24.

Earley, P.C. (1985), "The Influence of Information, Choice and Task Complexity upon Goal Acceptance, Performance and Personal Goals," *Journal of Applied Psychology* 70:3, 481–91.

Eggleston, R.G. (2003), "Work-centered Design: A Cognitive Engineering Approach to System Design," in *Proceedings of the Human Factors and Ergonomics Society 47th Annual Meeting* (Santa Monica, CA: Human Factors and Ergonomics Society) 263–7.

Eisenhardt, K.M., and Martin, J. (2000), "Dynamic Capabilities: What are They?" *Strategic Management Journal* 21:10–11, 1105–21.

Eisenhardt, K.M., and Tabrizi, B.N. (1995), "Accelerating Adaptive Processes: Product Innovation in the Global Computer Industry," *Administrative Science Quarterly* 40:1, 84–110.

Feltovich, P.J., Hoffman, R.R., Woods, D.D., and Roesler, A. (2004), "Keeping It Too Simple: How the Reductive Tendency Affects Cognitive Engineering," *IEE Intelligent Systems* May–June, 90–94.

FEMA (Federal Emergency Management Agency) (2005), *Help After a Disaster: Applicant's Guide to the Individuals & Households Program* (Washington, DC: FEMA).

FEMA (Federal Emergency Management Agency) (2008), *Mitigation*, 20 December 2007 <http://www.fema.gov/government/mitigation.shtm>, accessed 27 March 2008.

Ferrier, W.J. (2001), "Navigating the Competitive Landscape: The Drivers and Consequences of Competitive Aggressiveness," *Academy of Management Journal* 44:4, 858–77.

Ferrier, W.J., Smith, K.G., and Grimm, C.M. (1999), "The Role of Competitive Action in Market Share Erosion and Industry Dethronement: A Study of Industry Leaders and Challengers," *Academy of Management Journal* 42:4, 372–88.

Fiksel, J. (2003), "Designing Resilient, Sustainable Systems," *Environmental Science and Technology* 37:23, 5330–9.

Fiol, C.M., and Lyles, M.A. (1985), "Organizational Learning," *Academy of Management Review* 10:4, 803–13.

Fisk, D. (2004), "Engineering Complexity," *Interdisciplinary Science Reviews* 29:2, 151–61.

Flach, J.M. (2003), "For Those Condemned to Live in the Future," *Quality and Safety in Healthcare* 12:4, 311–12.

Folke, C. (2006), "Resilience: The Emergence of a Perspective for Social-Ecological Systems Analysis," *Global Environmental Change* 16:3, 253–67.

Folke, C., Hahn, T., Olsson, P., and Norberg, J. (2005), "Adaptive Governance of Social-ecological Systems," *Annual Review of Environmental Resources* 30:1, 441–73.

Fraser, J., Smith, P.J., and Smith, J. (1992), "A Catalog of Errors," *International Journal of Man-Machine Studies* 37:3, 265–307.

Freilich, M. (1991), "Smart Rules and Proper Rules: A Journey through Deviance," in M. Freilich, D. Raybeck and J. Savishinsky (eds), *Deviance: Anthropological Perspectives* (Westport, CT: Greenwood Publishing Group) 27–50.

Freitag, R. (2001), "The Impact of Project Impact on the Nisqually Earthquake," *Natural Hazards Observer* 25:5 (published online May 2001) <http://www.colorado.edu/hazards/o/archives/2001/may01/may01a.html#nisqually>.

Fujita Y. (2006), "Remedies," in E. Hollnagel, D.D.Woods and N. Levenson (eds), *Resilience Engineering* (Aldershot, UK: Ashgate).

Furedi, F. (2007), "The Changing Meaning of Disaster," *Area* 39:4, 482–9.

Gaba, D. (2003, Spring), *"Safety First: Ensuring Quality Care in the Intensely Productive Environment – The HRO Model,"* APSF *Newsletter.* Anesthesia Patient Safety Foundation.

Garg, A., Adhikari, N., McDonald, H., Rosas-Arellano, M., Deveraux, P., Beyene, J. et al. (2005), "Effects of Computerized Clinical Decision Making Support Systems on Practitioner Performance and Patient Outcomes," *JAMA* 293:10, 1223–38.

Garrod, S., Fay, N., Lee, J., Oberlander, J., and MacLeod, T. (2007), "Functions of Representation: Where Might Graphical Symbol Systems Come From?" *Cognitive Science* 31:6, 961–87.

Geschwind, C.H. (2001), *California Earthquakes: Science, Risk, and the Politics of Hazard Mitigation* (Baltimore: Johns Hopkins University Press).

Ghemawat, P., and del Sol, P. (1998), "Commitment Versus Flexibility," *California Management Review* 40:4, 26–42.

Glanz, J., Wong, E., and Broad, W.J. (2003), "Agency that Soared is Now Held Down By its Bureaucracy," *New York Times*, 18 February, A1, A16.

Godschalk, D.R. (2003), "Urban Hazard Mitigation: Creating Resilient Cities," *Natural Hazards Review* 4:3, 136–43.

Goggin, M.L., Bowman, A., Lester, J., and O'Toole, L. (1990), *Implementation Theory and Practice: Toward a Third Generation* (Glenview, IL: Scott Foresman/Little Brown).

Goldman, S.L., Nagel, R.N., and Preiss, K. (1995), *Agile Competitors and Virtual Organizations* (New York: John Wiley).

Green, J.J., Gill, D.A., and Kleiner, A.M. (2006), "From Vulnerability to Resiliency: Assessing Impacts and Responses to Disaster," *Southern Rural Sociology* 21:2, 89–99.

Griffey, R., Wittles, K., Gilboy, N., and McAfee, A.T. (forthcoming), "Use of a Computerized Forcing Function Improves Performance in Ordering Restraints," *Annals of Emergency Medicine*.

Grimm, C.M., Lee, H., and Smith, K.G. (2006), *Strategy as Action: Competitive Dynamics and Competitive Advantage* (Oxford: Oxford University Press).

Guerlain, S., Smith, P.J., Obradovich, J.H., Rudmann, S., Strohm, P., Smith, J.W., et al. (1999), "Interactive Critiquing as a Form of Decision Support: An Empirical Evaluation," *Human Factors* 41:1, 72–89.

Gunderson, L.H. (2000), "Ecological Resilience – In Theory and Application," *Annual Review of Ecology and Systematics* 31, 425–39.

Haller, A., Cimpian, E., Mocan, A., Oren, E., and Bussler, C. (2005), "WSMX-A Semantic Service-Oriented Architecture," *Proceedings of the 2005 International Conference on Web Services* 1, 321–8.

Hamel, G., and Prahalad, C.K. (1993), "Strategy as Stretch and Leverage," *Harvard Business Review* 71:2, 75–85.

Hamel, G., and Prahalad, C.K. (1994), *Competing for the Future* (Boston, MA: Harvard Business School Press).

Hamel, G., and Valikangas, L. (2003), "The Quest for Resilience," *Harvard Business Review* 81:9, 52–63.

Hammonds, K.H. (2002), "The Strategy of a Fighter Pilot," *Fast Company* 59, 98–105.

Harrald, J.R. (2006), "Agility and Discipline: Critical Success Factors for Disaster Response," *The Annals of the American Academy of Political and Social Science* 604:1, 256–72.

Helmreich R.L., and Merritt, A.C. (1998), *Culture at Work in Aviation and Medicine* (Aldershot, UK: Ashgate).

Hinsz, V.B., Tindale, R.S., and Vollrath, D.A. (1997), "The Emerging Conceptualization of Groups as Information Processors," *Psychological Bulletin* 121:1, 43–64.

Holling, C.S. (1973), "Resilience and Stability of Ecological Systems," *Annual Review of Ecological Systems* 4, 1–23.

Holling, C.S., Gunderson, L.H., and Ludwig, D. (2002), *Panarchy: Understanding Transformations in Human and Natural Systems* (Washington DC: Island Press).

Hollnagel, E. (1993), *Human Reliability Analysis: Context and Control* (London: Academic Press).

Hollnagel, E. (1998), *Cognitive Reliability and Error Analysis Method* (London: Elsevier).

Hollnagel, E. (2004), *Barriers and Accident Prevention* (Aldershot, UK: Ashgate).

Hollnagel, E. (2006), "Resilience – The Challenge of the Unstable," in E. Hollnagel, D. Woods, and N. Leveson (eds), *Resilience Engineering – Concepts and Precepts* (Burlington, VT: Ashgate) 9–17.

Hollnagel, E. (2008a), "Investigation as an Impediment to Learning," in E. Hollnagel, C.P. Nemeth, and S. Dekker (eds), *Remaining Sensitive to the Possibility of Failure* (Aldershot, UK: Ashgate) 259–68.

Hollnagel, E. (2008b), "From Protection to Resilience: Changing Views on How to Achieve Safety," in *Proceedings of the 8th International Symposium of the Australian Aviation Psychology Association*, 8–11 April, Sydney, Australia.

Hollnagel, E. (forthcoming), *The Four Qualities of Resilience, in Preparation and Restoration: Resilience in Human Systems* (Burlington, VT: Ashgate).

Hollnagel, E., Nemeth, C., and Dekker, S. (2008), *Remaining Sensitive to the Possibility of Failure*. Resilience Engineering Perspectives, Vol. 1 (Aldershot, UK: Ashgate).

Hollnagel, E., and Sundstrom, G. (2006), "States of Resilience," in E. Hollnagel, D.D. Woods, and N. Leveson (eds), *Resilience Engineering: Concepts and Precepts* (Aldershot, UK: Ashgate).

Hollnagel, E., and Woods, D. (2005), *Joint Cognitive Systems: Foundations of Cognitive Systems Engineering* (Boca Raton, FL: Taylor and Francis/ CRC Press).

Hollnagel, E., and Woods, D. (2006), "Epilogue: Resilience. Engineering Precepts," in E. Hollnagel, D.D. Woods, and N. Leveson (eds), *Resilience Engineering: Concepts and Precepts* (Aldershot, UK: Ashgate).

Hollnagel, E., Woods, D., and Leveson, N. (2004), "About Resilience Engineering..." *International Symposium on Resilience Engineering*. Söderköping, Sweden. October <http://csel.eng.ohio-state.edu/ woods/error/About%20Resilience%20Engineer.pdf>, accessed 14 July 2008.

Hollnagel, E., Woods, D.D., and Leveson, N. (eds) (2006), *Resilience Engineering: Concepts and Precepts* (Aldershot, UK: Ashgate).

Hong, L., and Page, S.E. (2002), "Groups of Diverse Problem Solvers Can Outperform Groups of High-ability Problem Solvers," *Proceedings of the National Academy of Sciences* 101:46, 16385–9.

Houston Chronicle (1995), "Sleight of Ham; Pork-free Disaster-aid Measures are Long Overdue," *Houston Chronicle*, 25 March 1995, 34.

Hunt-Smith, J., Donaghy, A., Leslie, K., Kluger, M., Gunn, K., and Warwick, N. (1999), "Safety and Efficacy of Target Controlled Infusion (Diprifusor) vs Manually Controlled Infusion of Propofol for Anesthesia," *Anaesthesia Intensive Care* 27:3, 260–4.

Hutchins, E. (2000), *Cognition in the Wild* (Cambridge, MA: MIT Press).

Hutchins, E. (2002), "Cognitive Artifacts," in R.A. Wilson and F.C. Keil (eds), *The MIT Encyclopedia of the Cognitive Sciences* (Cambridge, MA: The MIT Press) <http://cognet.mit.edu/library/erefs/mitecs/hutchins.html>.

Inkpen, A.C., and Tsang, E.W.K. (2005), "Social Capital, Networks, and Knowledge Transfer," *Academy of Management Review* 30:1, 146–65.

INSAG (International Nuclear Safety Group) (1995), *Defence in Depth in Nuclear Safety* (INSAG-10) (Vienna: International Atomic Energy Agency).

Ireland, R.D., Hitt, M.A., and Vaidyanath, D. (2002), "Alliance Management as a Source of Competitive Advantage," *Journal of Management* 28:3, 413–46.

Jackman, W.J., Russell, T.H., and Chanute, O. (1910), *Flying Machines: Construction and Operation* (Chicago: Charles C. Thompson Co.).

Jacobs, B. (2005), "Urban Vulnerability: Public Management in a Changing World," *Journal of Contingencies & Crisis Management* 13:2, 39–43.

Jamrog, J.J., McCann, J.E., Lee, J.M., Morrison, C.L., Selsky, J.W., and Vickers, M. (2006), *Agility and Resilience in the Face of Continuous Change: A Global Study of Current Trends and Future Possibilities 2006-2016* (New York: American Management Association).

Judge, W.Q., Fryxell, G.E., and Dooley, R.S. (1997), "The New Task for R&D Management: Creating Goal-directed Communities for Innovation," *California Management Review* 39:3, 72–86.

Kaplan, B. (2001), "Evaluating Informatics Applications – Clinical Decision Support Systems Literature Review," *International Journal of Biomedical Informatics* 64:1, 15–37.

Kaufman, W., and Pilkey, O.H. (1983), *The Beaches are Moving: the Drowning of America's Shoreline* (Durham, NC: Duke University Press).

Kelly, T.J. (2001), *Moon Lander: How We Developed the Apollo Lunar Module* (Washington, DC: Smithsonian Institute).

Kendra, J.M., and Wachtendorf, T. (2003), "Elements of Resilience after the World Trade Center Disaster: Reconstructing New York City's Emergency Operations Center," *Disasters* 27:1, 37–53.

Kendra, J.M., and Wachtendorf, T. (2006), "Community Innovation and Disasters," in H. Rodríguez, E.L. Quarantelli, and R.R. Dynes, *Handbook of Disaster Research* (New York: Springer).

Kerr, D.S., and Murthy, U.S. (2004), "Divergent and Convergent Idea Generation in Teams: A Comparison of Computer-Mediated and Face-to-Face Communication," *Group Decision and Negotiation* 13:4, 381–99.

Kerr, N.L., and Tindale, R.S. (2004), "Group Performance and Decision Making," *Annual Review of Psychology* 55:1, 623–55.

Kintisch, E. (2005), "HURRICANE KATRINA: Levees Came Up Short, Researchers Tell Congress," *Science* 310:5750, 953–5.

Kirton, M. (1976), "Adaptors and Innovators: A Description and Measure," *Journal of Applied Psychology* 61:5, 622–9.

Klein, G., Woods, D., Bradshaw, J.M., Hoffman, R.R., and Feltovich, P.J. (2004), "Ten Challenges for Making Automation a Team Player in Joint Human-agent Activity," *IEEE Computer* 19:6, 91–5.

Kogut, B., and Zander, U. (1996), "What Firms Do? Coordination, Identity, and Learning," *Organization Science* 7:5, 502–18.

Kranz, G. (2000), *Failure is Not an Option: Mission Control from Mercury to Apollo 13 and Beyond* (New York: Simon and Schuster).

Kreps, G.A. (1984), "Sociological Inquiry and Disaster Research," *Annual Review of Sociology* 10, 309–30.

Langewiesche, W. (2002), *American Ground, Unbuilding the World Trade Center* (New York: North Point Press).

Lasswell, H.D. (1958), *Politics: Who Gets What, When, How* (New York: Meridian Books).

Lautman, L., and Gallimore, P.L. (1987), "Control of the Crew Caused Accident: Results of a 12-operator Survey," *Boeing Airliner*, April–June, 1–6.

Leape, L.L., Woods, D.D., Halie, M.J., Kizer, K.W., Schroeder, S.A., and Lundberg, G.D. (1998), "Promoting Patient Safety by Preventing Medical Errors," *JAMA* 280:16, 1444–7.

Lei, D., Hitt, M.A., and Bettis, R. (1996), "Dynamic Core Competences through Meta-Learning and Strategic Context," *Journal of Management* 22:4, 549–69.

Lengnick-Hall, C.A., and Beck, T.E. (2003), "Beyond Bouncing Back: The Concept of Organizational Resilience," Paper presented at the National Academy of Management meetings, Seattle, WA.

Lengnick-Hall, C.A., and Beck, T.E. (2005), "Adaptive Fit versus Robust Transformation: How Organizations Respond to Environmental Change," *Journal of Management* 31:5, 738–57.

Lengnick-Hall, M.L., and Lengnick-Hall, C.A. (2003), *Human Resource Management in the Knowledge Economy: New Challenges, New Roles, New Capabilities* (San Francisco, CA: Berrett-Koehler Publishers).

Leveson, N. (2003), "A New Accident Model for Engineering Safer Systems," *Safety Science* 42:4, 237–70.

Leveson, N., Dulac, N., Zipkin, D., Cutcher-Gershenfeld, J., Carroll, J., and Barrett, B. (2006), "Engineering Resilience into Safety-Critical Systems," in E. Hollnagel, D. Woods, and N. Leveson (eds), *Resilience Engineering: Concepts and Precepts* (Burlington, VT: Ashgate).

Lewis, D., and Mioch, J. (2005), "Urban Vulnerability and Good Governance," *Journal of Contingencies & Crisis Management* 13:2, 50–3.

Logan, J.R., and Molotch, H.L. (1988), *Urban Fortunes: The Political Economy of Space* (Berkeley: University of California Press).

Logsdon, J. (ed.) (1999), *Managing the Moon Program: Lessons Learned from Project Apollo*. Monographs in Aerospace History. Number 14, July (Washington, DC: NASA).

Lovell, J., and Kluger, J. (1995), *Apollo 13* (New York: Simon and Schuster).

Luff, P., Heath, C., and Greatbatch, D. (1992), "Task-in-interaction: Paper and Screen Based Documentation in Collaborative Activity," in *Proceedings of ACM CSCW' 92 Conference on Computer-Supported Cooperative Work* (New York: Association for Computing Machinery).

Mallak, L.A. (1998a), "Measuring Resilience in Health Care Provider Organizations," *Health Manpower Management* 24:4, 148–52.

Mallak, L.A. (1998b), "Putting Organizational Resilience to Work," *Industrial Management* 40:6, 8–13.

Manyena, S.B. (2006), "The Concept of Resilience Revisited," *Disasters* 30:4, 434–50.

Marais, K.B., and Saleh, J.H. (2008), "Conceptualizing and Communicating Organizational Risk Dynamics in the Thoroughness-efficiency Space," *Reliability Engineering and Systems Safety* 93:11, 1710–9.

March, J.G. (1991), "Exploration and Exploitation in Organizational Learning," *Organization Science* 2:1, 71–87.

March, J.G., and Levinthal, D.A. (1993), "The Myopia of Learning," *Strategic Management Journal* 14 (special issue): 95–112.

May, P.J. (1993), "Mandate Design and Implementation: Enhancing Implementation Efforts and Shaping Regulatory Styles," *Journal of Policy Analysis and Management* 10:2, 634–63.

May, P.J. (1994), "Analyzing Mandate Design: State Mandates Governing Hazard-Prone Areas," *Publius: The Journal of Federalism* 24:2, 1–15.

May, P.J. (1995), "Can Cooperation be Mandated? Implementing Intergovernmental Environmental Management in New South Wales and New Zealand," *Publius: The Journal of Federalism* 25:1, 89–113.

May, P.J., and Birkland, T.A. (1994), "Earthquake Risk Reduction: An Examination of Local Regulatory Efforts," *Environmental Management* 18:6, 923–39.

McCann, J. (2004), "Organizational Effectiveness: Changing Concepts for Changing Environments," *Human Resource Planning* 27:1, 42–50.

McCurdy, H.E. (1993), *Inside NASA: High Technology and Organizational Change in the U.S. Space Program* (Baltimore: Johns Hopkins).

MCEER (Multidisciplinary Center for Earthquake Engineering Research) (2006), *MCEER's Resilience Framework* (Buffalo, NY: MCEER, University at Buffalo, State University of New York).

McGregor, D. (1966), *Leadership and Motivation: The Essays of Douglas McGregor* (Cambridge, MA: MIT Press).

McNutt, R., Abrams, R., and Hasler, S. (2005), "Diagnosing Diagnostic Mistakes," AHRQ Web M&M (published online May 2005) <http: www.webmm.ahrq.gov/printview.aspx?caseID=95>.

Mendonça, D. (2007), "Decision Support for Improvisation in Response to Extreme Events," *Decision Support Systems* 43:3, 952–67.

Mendonça, D., Beroggi, G.E.G., van Gent, D., and Wallace, W.A. (2006), "Assessing Group Decision Support Systems for Emergency Response Using Gaming Simulation," *Safety Science* 44:6, 523–35.

Mendonça, D., and Wallace, W.A. (2004), "Studying Organizationally-situated Improvisation in Response to Extreme Events," *International Journal of Mass Emergencies and Disasters* 22:2, 5–29.

Mendonça, D., and Wallace, W.A. (2007), "A Cognitive Model of Improvisation in Emergency Management," *IEEE Transactions on Systems, Man, and Cybernetics: Part A* 37:4, 547–61.

*Merriam-Webster Collegiate Dictionary* (2003), (Springfield MA: Merriam Webster, Inc.).

Meyer, A.D. (1982), "Adapting to Environmental Jolts," *Administrative Science Quarterly* 27:4, 515–37.

Mileti, D.S. (1999), *Disasters by Design: A Reassessment of Natural Hazards in the United States* (Washington, DC: Joseph Henry Press).

Mittler, E. (1997), *A Case Study of Florida's Emergency Management Since Hurricane Andrew*. Natural Hazards Research Working Paper #98 (Boulder: Natural Hazard Research and Applications Information Center, University of Colorado).

Morgan, G. (1997), *Images of Organization*, 2nd edn (Thousand Oaks, CA: Sage Publications).

Morgenstern, J. (1995), "The Fifty-nine Story Crisis," *The New Yorker* 71:14, 45–53.

Morris, P.W.G. (1994), *The Management of Projects* (London: Thomas Telford).

Multihazard Mitigation Council (2005), *Natural Hazard Mitigation Saves: An Independent Study to Assess the Future Savings from Mitigation Activities* (Washington, DC: National Institute of Building Sciences) <http://www.nibs.org/MMC/mmcactiv5.html>.

Mumaw, R.J., Roth, E.M., Vicente, K.J., and Burns, C.M. (2000), "There is More to Monitoring a Nuclear Power Plant than Meets the Eye," *Human Factors* 42:1, 36–55.

Murray, M. (2001), "A Tough Act to Follow at FEMA," *National Journal* 33:22, 1664–5.

Mustafa, D. (2003), "Reinforcing Vulnerability? Disaster Relief, Recovery, and Response to the 2001 Flood in Rawalpindi, Pakistan," *Global Environmental Change Part B: Environmental Hazards* 5:3-4, 71–82.

Nathanael, D., and Marmaras, N. (2008), "Work Practices and Prescriptions: A Key Issue for Organizational Resilience," in E. Hollnagel, C.P. Nemeth, and S. Dekker (eds), *Remaining Sensitive to the Possibility of Failure* (Aldershot, UK: Ashgate) 101–18.

Nemeth, C. (2005), "Health Care Forensics," in G. Salvendy (ed.), *Handbook of Human Factors in Litigation* (New York: Taylor and Francis) 37–1 to 37–18.

Nemeth, C. (2007), "Healthcare Groups at Work: Further Lessons from Research into Large Scale Coordination," *Cognition, Technology and Work* 9:3, 127–30.

Nemeth, C.P., Cook, R.I., and Woods, D.D. (2004), "The Messy Details: Insights from the Study of Technical Work in Health Care," *IEEE Transactions on Systems, Man and Cybernetics: Part A* 34:6, 689–92.

Nemeth, C., Nunnally, M., O'Connor, M., and Cook, R. (2006), "Creating Resilient IT: How the Sign-out Sheet Shows Clinicians Make Healthcare Work," in *Proceedings of the American Medical Informatics Association Annual Symposium*. Washington, DC: American Medical Informatics Association.

Nemeth, C., Nunnally, M., O'Connor, M., Brandwijk, M., Kowalsky, J., and Cook, R. (2007), "Regularly Irregular: How Groups Reconcile Cross-cutting Agendas in Healthcare," *Cognition, Technology and Work* 9:3, 139–48.

Nemeth, C., Nunnally, M., O'Connor, M., Klock, P.A., and Cook, R. (2005), "Getting to the Point: Developing IT for the Sharp End of Healthcare," *Journal of Biomedical Informatics* 38:1, 18–25.

Nemeth, C., Wears, R., Woods, D., Hollnagel, E., and Cook, R. (2008), "Minding the Gaps: Creating Resilience in Healthcare," in K. Henriksen, J.B. Battles, M.A. Keyes, and M.L. Grady (eds), *Advances in Patient Safety: New Directions and Alternative Approaches. Vol. 3. Performance and Tools*. AHRQ Publication No. 08-0034-3 (Rockville, MD: Agency for Healthcare Research and Quality) 259–71.

Neufeld, M.J. (2007), *Von Braun: Dreamer of Space, Engineer of War* (New York: Knopf).

Norris, G. (2007), "Small Wonder: F-22 Drops its First SDB, But Not Without Test Incident," *Aviation Week & Space Technology* 167:14, 35.

North Carolina Division of Emergency Management (2004), *State 322 Natural Hazard Mitigation Plan* (Raleigh, NC: North Carolina Division of Emergency Management).

North Carolina Office of the Governor, and North Carolina Office of State Budget and Management (2006), *North Carolina Disaster Recovery Guide* (Raleigh, NC: NC Office of the Governor).

Nunnally, M., Nemeth, C., Brunetti, V., and Cook, R. (2004), "Lost in Menuspace: User Interactions with Complex Medical Devices," *IEEE Transactions on Systems, Man and Cybernetics-Part A* 34:6, 736–42.

Obradovich, J.H., and Smith, P.J. (2008), "Design Concepts for Virtual Work Systems," in J. Nemiro, M. Beyerlein, L. Bradley, and S. Beyerlein (eds), *The Handbook of High-Performance Virtual Teams* (San Francisco: Jossey-Bass) 9–17.

Orasanu, J., Martin, L., and Davison, J. (2001), "Cognitive and Contextual Factors in Aviation Accidents: Decision Errors," in E. Salas and G. Klein (eds), *Linking Expertise and Naturalistic Decision Making* (Mahwah, NJ: Lawrence Erlbaum Associates).

O'Reilly, C.A.I., and Tushman, M.L. (2004), "The Ambidextrous Organization," *Harvard Business Review* 82:4, 74–81.

Orlady, H.W., and Orlady, L.M. (1999), *Human Factors in Multi-Crew Flight Operations* (Aldershot, UK: Ashgate).

Ostrom, E. (1999), "Coping with Tragedies of the Commons," *Annual Review of Political Science* 2, 493–535.

*Oxford English Dictionary, ed. II* (1989), (Oxford: Clarendon Press).

Papazoglou, M.P. (2003), "Service-oriented Computing: Concepts, Characteristics, and Directions," in *Proceedings of the Fourth International Conference on Web Information Systems Engineering* (Rome: IEEE Computer Society) 3–12.

Patterson, E.S., Watts-Perotti, J.C., and Woods, D.D. (1999), "Voice Loops as Coordination Aids in Space Shuttle Mission Control," *Computer Supported Cooperative Work* 8:4, 353–71.

Perrey, R., and Lycett, M. (2003), "Service-oriented Architecture," in *Proceedings of the Symposium on Applications and the Internet Workshops* (Washington, DC: IEEE Computer Society) 116–19.

Perrow, C. (1984), *Normal Accidents* (New York: Basic Books).

Perrow, C. (1999), *Normal Accidents: Living with High-Risk Technologies* (Princeton, NJ: Princeton University Press).

Perry, S., McDonald, S., Anderson, B., Tran, T., and Wears, R. (2007a), "Ironies of Improvement: Organizational Factors Undermining Resilient Performance in Healthcare," *IEEE International Conference on Systems, Cybernetics and Man* (Montreal, Canada) 3413–17.

Perry, S., Wears, R.L., Anderson, B., and Booth, A. (2007b), "Peace and War: Contrasting Cases of Resilient Teamwork in Healthcare." Paper presented at the 8th International Naturalistic Decision-making Conference Proceedings, Pacific Grove, CA.

Petroski, H. (1992), *To Engineer is Human: The Role of Failure in Successful Design* (New York: Vintage Books).

Pilkey, O.H., and Dixon, K.L. (1998), *The Corps and the Shore* (Washington, DC: Island Press).

Pimm, S.L., and Lawton, J.H. (1977), "Number of Trophic Levels in Ecological Communities," *Nature* 268, 329–31.

Platt, R.H. (1999), *Disasters and Democracy* (Washington, DC: Island Press).

Prater, C.S., and Lindell, M.K. (2000), "Politics of Hazard Mitigation," *Natural Hazards Review* 1:2, 73–82.

Rasmussen, J. (1983), Position paper for NATO Conference on Human Error. Bellagio, IT.

Rasmussen, J. (1997a), "Merging Paradigms: Decision Making, Management, and Cognitive Control," in R. Flin, E. Salas, M. Strub, and L. Martin (eds), *Decision Making Under Stress: Emerging Themes and Applications* (Aldershot, UK: Ashgate) 67–81.

Rasmussen, J. (1997b), "Risk Management in a Dynamic Society: A Modeling Problem," *Safety Science* 27:2/3, 183–213.

Reason, J. (1991), *Human Error* (Cambridge, UK: Cambridge Press).

Reese, S. (2005), *Risk-Based Funding in Homeland Security Grant Legislation: Analysis of Issues for the 109th Congress* (Washington, DC: Congressional Research Service). <http://fas.org/sgp/crs/homesec/RL33050.pdf>, accessed 20 March 2008.

Reid, P.R., Compton, W.D., Grossman, J.H., and Fanjiang, G. (eds) (2005), *Building a Better Delivery System: A New Engineering/Health Care Partnership* (Washington, DC: The National Academies Press).

Rindova, V.P., and Kotha, S. (2001), "Continuous 'Morphing': Competing through Dynamic Capabilities, Form, and Function," *Academy of Management Journal* 44:6, 1263–80.

Rochlin, G.I. (1999), "Safe Operation as a Social Construct," *Ergonomics* 42:11, 1549–60.

Rochlin, G.I., La Porte, T.R., and Roberts, K.H. (1987), "The Self-designing High-Reliability Organization, Aircraft Carrier Flight Operations at Sea," *Naval War College Review* 40:4, 76–90.

Rognin, L., Salembier, P., and Zouinar, M. (2000), "Cooperation, Reliability of Socio-technical Systems and Allocation of Function," *International Journal of Human-Computer Studies* 52:2, 357–79.

Roth, A.E. (2008), "What Have We Learned from Market Design?" *Economic Journal* 118:527, 285–310.

Roth, A.V. (1996), "Achieving Strategic Agility through Economies of Knowledge," *Strategy & Leadership* 24:2, 30–7.

Roth, E.M., Malsch, N., and Multer, J. (2001), *Understanding How Train Dispatchers Manage and Control Trains: Results of a Cognitive Task Analysis* (Washington, DC: US Department of Transportation/Federal Railroad Administration).

Roth, E.M., Malsch, N., Multer, J., and Coplen, M. (1999), "Understading How Train Dispatchers Manage and Control Trains: A Cognitive Analysis of a Distributed Planning Task," in *Proceedings of the Human Factors and Ergonomics Society 43rd Annual Meeting* (Santa Monica, CA: Human Factors and Ergonomics Society).

Roth, E.M., Multer, J., and Raslear, T. (2006), "Shared Situation Awareness as a Contributor to High Reliability Performance in Railroad Operations," *Organization Studies* 27:7, 967–87.

Roth, E.M., and Patterson, E.S. (2005), "Using Observational Study as a Tool for Discovery: Uncovering Cognitive and Collaborative Demands and Adaptive Strategies," in H. Montgomery, R. Lipshitz, and B. Brehmer (eds), *How Professionals Make Decisions* (Mahwah, NJ: Lawrence Erlbaum Associates).

Roth, E.M., Scott, R., Deutsch, S., Kuper, S., Schmidt, V., Stilson, M., et al. (2006), "Evolvable Work-centered Support Systems for Command and Control: Creating Systems Users Can Adapt to Meet Changing Demands," *Ergonomics* 49:7, 688–705.

Roth, E., Scott, R., Whitaker, R., Kazmierczak, T., Forsythe, M., Thomas, G., et al. (2007), "Designing Decision Support for Mission Resource Retasking," in *Proceedings of the 2007 International Symposium on Aviation Psychology* (Atlanta, GA: Curran Associates).

Roth, E.M., Stilson, M., Scott, R., Whitaker, R., Kazmierczak, T., Thomas-Meyers, G., et al. (2006), "Work-centered Design and Evaluation of a C2 Visualization Aid," in *Proceedings of the Human Factors and Ergonomics Society 49th Annual Meeting* (Santa Monica, CA: Human Factors and Ergonomics Society).

Salamon, L.M., and Lund, M.S. (1989), "The Tools Approach: Basic Analytics," in M.S. Lund and L.M. Salamon (eds), *Beyond Privatization: The Tools of Government Action* (Washington, DC: Urban Institute Press).

Salas, E., Dickinson, T.L., Converse, S.A., and Tannenbaum, S.I. (1992), "Toward an Understanding of Team Performance and Training," in R.W. Sweezey and E. Salas (eds), *Teams: Their Training and Performance* (Norwood, NJ: Ablex Publishing).

Santos-Reyes, J., and Beard, A.N. (2006), "A Systematic Analysis of the Paddington Railway Accident," *Proc. IMechE Part F: J. Rail and Rapid Transit* 220:2, 121–51.

Sarter, N., Woods, D., and Billings, C. (1997), "Automation Surprises," in G. Salvendy (ed.), *Handbook of Human Factors and Ergonomics* (New York: John Wiley and Sons) 1926–43.

Scanlon, J. (1994), "The Role of EOCs in Emergency Management: A Comparison of American and Canadian Experience," *International Journal of Mass Emergencies and Disasters* 12:1, 51–75.

Scavo, C., Kearney, R.C., and Kilroy, R.J. (2008), "Challenges to Federalism: Homeland Security and Disaster Response," *Publius: The Journal of Federalism* 38:1, 81–110.

Scheffer, M., Hosper, S.H., Meijer, M.L., and Moss, B. (1993), "Alternative Equilibria in Shallow Lakes," *Trends in Evolutionary Ecology* 8:8, 275–9.

Schulman, P.R. (1993a), "Analysis of High Reliability Organizations: A Comparative Framework," in K.H. Roberts (ed.), *New Challenges to Understanding Organizations* (New York: Macmillan).

Schulman, P.R. (1993b), "The Negotiated Order of Organizational Reliability," *Administration & Society* 25:3, 353–72.

Scott, R., Roth, E.M., Deutsch, S., Malchiodi, E., Kazmierczak, T., Eggleston, R.G. et al. (2005), "Work-centered Support Systems: A Human-centered Approach to Intelligent System Design," *IEEE Intelligent Systems* 20:2, 73–81.

Seed, R.B., Bea, R.G., Abdelmalak, R.I., Athanasopoulos, A.G., Boutwell, G.P., Bray, J.D., et al. (2006), *Investigation of the Performance of the New Orleans Flood Protection Systems in Hurricane Katrina on August 29, 2005* (Berkeley, CA: Independent Levee Investigation Team, University of California, Berkeley). <http://www.ce.berkeley.edu/~new_orleans/report/intro&summary.pdf>, accessed 25 January 2008.

Senge, P.M., Roberts, C., Ross, R.B., Smith, B.J., and Kleiner, A. (1994), *The Fifth Discipline Fieldbook: Strategies and Tools for Building a Learning Organization* (New York: Free Press).

Sharpe, V., and Faden, A. (1998), *Medical Harm* (Cambridge, UK: Cambridge University Press) 214–20.

Sheremata, W.A. (2000), "Centrifugal and Centripetal Forces in Radical New Product Development under Time Pressure," *Academy of Management Review* 25:2, 389–408.

Shockley, W.B. (1974), "The Invention of the Transistor, an Example of Creative-failure Methodology," in *The Public Need and the Role of the Inventor* (Washington: National Bureau of Standards).

Simon, H.A. (1962), "The Architecture of Complexity," in *Proceedings of the American Philosophical Society* 106:6, 467–82.

Simons, A.J. (1997), *The Company They Keep: Life Inside the U.S. Army Special Forces* (New York: Avon).

Smith, K.G., Ferrier, W.J., and Grimm, C.M. (2001), "King of the Hill: Dethroning the Industry Leader," *Academy of Management Executive* 15:2, 59–70.

Smith, P.J., Beatty, R., Spencer, A., and Billings, C. (2003), "Dealing with the Challenges of Distributed Planning in a Stochastic Environment: Coordinated Contingency Planning," in *Proceedings of the 2003 Annual Conference on Digital Avionics Systems* (Chicago, IL: IEEE).

Smith, P.J., Bennett, K., and Stone, B. (2006), "Representation Aiding to Support Performance on Problem Solving Tasks," in R. Williges (ed.), *Reviews of Human Factors and Ergonomics*, Volume 2 (Santa Monica, CA: Human Factors and Ergonomics Society).

Smith, P.J., Geddes, N., and Beatty, R. (2008), "Human-centered Design of Decision Support Systems," in A. Sears and J. Jacko (eds), *Handbook of Human-Computer Interaction*, 2nd edn (Mahwah, NJ: Lawrence Erlbaum Associates).

Smith, P.J., Giffin, W., Rockwell, T., and Thomas, M. (1986), "Modeling Fault Diagnosis as the Activation and Use of a FrameSystem," *Human Factors* 28:6, 703–16.

Smith, P.J., McCoy, E., and Layton, C. (1997), "Brittleness in the Design of Cooperative Problem-solving Systems: The Effects on User Performance," *IEEE Transactions on Systems, Man and Cybernetics* 27:3, 360–71.

Smith, P.J., McCoy, E., and Orasanu, J. (2001), "Distributed Cooperative Problem-solving in the Air Traffic Management System," in E. Salas and G. Klein (eds), *Linking Expertise and Naturalistic Decision Making* (Mahwah, NJ: Laurence Erlbaum Associates) 369–84.

Smith, P.J., and Rudmann, S. (2005), "Clinical Decision Making and Diagnosis: Implications for Immunohematologic Problem-solving," in S. Rudmann (ed.), *Serologic Problem-Solving: A Systematic Approach for Improved Practice* (Bethesda: AABB Press) 1–16.

Smith, P.J., Spencer, A.L., and Billings, C. (2007), "Strategies for Designing Distributed Systems: Case Studies in the Design of an Air Traffic Management System," *Cognition, Technology and Work* 9:1, 39–49.

Smith, P.J., Stone, R.B., and Spencer, A. (2006), "Design as a Prediction Task: Applying Cognitive Psychology to System Development," in W. Marras and W. Karwowski (eds), *Handbook of Industrial Ergonomics*, 2nd edn (New York: Marcel Dekker Inc.) 24-1–24-18.

Squires, A. (1986), *The Tender Ship: Government Management of Technological Change* (Boston: Birkhauser).

Stasser, G. (1999), "A Primer of Social Decision Scheme Theory: Models of Group Influence, Competitive Model-Testing, and Prospective Modeling," *Organizational Behavior and Human Decision Processes* 80:1, 3–20.

Stromberg, S.P., and Carlson, J. (2006), "Robustness and Fragility in Immunosenescence," *Computational Biology* 2:11, 1475–81.

Suchman, L.A. (1987), *Plans and Situated Actions: The Problem of Human-machine Communication* (Cambridge, UK: Cambridge University Press).

Sutcliffe, K.M., and Vogus, T.J. (2003), "Organizing for Resilience," in K.S. Cameron, J.E. Dutton, and R.E. Quinn (eds), *Positive Organizational Scholarship: Foundations of a New Discipline* (San Francisco: Berrett-Koehler) 94–110.

Teece, D.J., Pisano, G., and Shuen, A. (1997), "Dynamic Capabilities and Strategic Management," *Strategic Management Journal* 18:7, 509–33.

Thomas, J.B., Clark, S.M., and Gioia, D.A. (1993), "Strategic Sensemaking and Organizational Performance: Linkages among Scanning, Interpretation, Action, and Outcomes," *Academy of Management Journal* 36:2, 239–70.

Thomas, L.G.I. (1996), "The Two Faces of Competition: Dynamic Resourcefulness and the Hypercompetitive Shift," *Organization Science* 7:3, 221–42.

Timmermans, S., and Berg, M. (2003), *The Gold Standard: The Challenge of Evidence-Based Medicine and Standardization in Health Care* (Philadelphia, PA: Temple University Press).

Tompkins, P.K. (1993), *Organizational Communication Imperatives: The Lessons of the Space Program* (Los Angeles: Roxbury).

Tompkins, P.K. (2005), *Apollo, Challenger, Columbia: The Decline of the Space Program* (Los Angeles: Roxbury).

Tripsas, M., and Gavetti, G. (2000), "Capabilities, Cognition, and Inertia: Evidence from Digital Imaging," *Strategic Management Journal* 21:10/11, 1147–61.

Tucker, A.L., and Edmondson, A.C. (2003), "Why Hospitals Don't Learn from Failures: Organizational and Psychological Dynamics that Inhibit System Change," *California Management Review* 45:2, 55–72.

Tversky, A., and Kahneman, D. (1974), "Judgment under Uncertainty: Heuristics and Biases," *Science* 185:4157, 1124–31.

US Dept. of Transportation (2000), *2000 Emergency Response Guidebook* (Washington DC: US Dept. of Transportation, Research and Special Programs Administration; Transport Canada, Safety and Security, Dangerous Goods; Secretariat of Transport and Communications).

van Heerden, I. (2007), "The Failure of the New Orleans Levee System Following Hurricane Katrina and the Pathway Forward," *Public Administration Review* 67:S1, 24–35.

Vicente, K.J. (1999), *Cognitive Work Analysis: Toward Safe, Productive, and Healthy Computer-Based Work* (Mahwah, NJ: Lawrence Erlbaum Associates).

Von Braun, W. (1956), "Teamwork: Key to Success in Guided Missiles," *Missiles and Rockets*, 1 (October), 38–43.

Wakin, D.J. (2001), "THE CRASH OF FLIGHT 587: THE RESPONSE; New Crisis, but This Time in Backyards of Rescuers," *New York Times*, 13 November.

Wald, M.L., and Schwartz, J. (2003), "Alerts Lacking, Shuttle Manager Says," *New York Times*, 23 July.

Walker, B., Holling, C.S., Carpenter, S.R., and Kinzig, A. (2004), "Resilience, Adaptability and Transformation in Social-Ecological Systems," *Ecology and Society* 9:2, 5 <http://www.ecologyandsociety.org/vol9/iss2/art5>.

Waterman, R.H. (1987), *The Renewal Factor* (New York: Bantam Books).

Watts-Perotti, J., and Woods, D.D. (2007), "How Anomaly Response is Distributed Across Functionally Distinct Teams in Space Shuttle Mission Control," *Journal of Cognitive Engineering and Decision Making* 1:4, 405–33.

Watts-Perotti, J., and Woods, D.D. (forthcoming), "Cooperative Advocacy: A Strategy for Integrating Diverse Perspectives in Anomaly Response," *Computer Supported Cooperative Work*.

Wears, R., and Berg, M. (2005), "Computers and Clinical Work-Reply. Letters," *JAMA* 294:2, 182.

Wears, R.L., Perry, S.J., McDonald, S., and Eisenberg, E. (2008), "Notes from Underground: The Role of Informal Communication Networks in Disjoint Understandings of Safety in Healthcare Organizations." Paper presented at the Human Factors in Organizational Design and Management, 9th International Symposium on Human Factors in Organizational Design and Management. Sao Paulo, Brazil, 19-21 March.

Wears, R.L., Perry, S.J., Anders, S., and Woods, D.D. (2008), "Resilience in the Emergency Department," in E. Hollnagel, C. Nemeth, and S.W.A. Dekker (eds), *Resilience Engineering Perspectives 1: Remaining Sensitive to the Possibility of Failure* (Aldershot, UK: Ashgate) 193–209.

Wears, R. L. and Woods, D. D. (2007). "Always Adapting," *Annals of Emergency Medicine* 50:5, 517–19.

Weick, K.E. (1988), "Enacted Sensemaking in Crisis Situations," *Journal of Management Studies* 25:4, 305–17.

Weick, K.E. (1993), "The Collapse of Sensemaking in Organizations: The Mann Gulch Disaster," *Administrative Science Quarterly* 38:4, 628–52.

Weick, K.E. (1995), *Sensemaking in Organizations* (Thousand Oaks, CA: Sage).

Weick, K.E., and Sutcliffe, K.M. (2001), *Managing the Unexpected: Assuring High Performance in an Age of Complexity* (San Francisco: Jossey-Bass).

Weick, K.E., and Sutcliffe, K.M. (2007), *Managing the Unexpected*, 2nd edn (San Francisco: Jossey-Bass).

Weick, K.E., Sutcliffe, K.M., and Obstfeld, D. (1999), "Organizing for High Reliability: Processes of Collective Mindfulness," in B.M. Staw and L.L. Cummings (eds), *Research in Organizational Behavior, Vol. 21* (Greenwich, CT: JAI Press) 81–123.

Weinberg, D.B. (2003), *Code Green: Money-Driven Hospitals and the Dismantling of Nursing* (Ithaca, NY: Cornell University Press).

Werner, E.E., and Smith, R.S. (2001), *Journeys from Childhood to Midlife: Risk, Resilience, and Recovery* (Ithaca, NY: Cornell University Press).

Westrum, R. (1993), "Cultures with Requisite Imagination," in J.A. Wise, V.D. Hopkin, and P. Stager (eds), *Verification and Validation of Complex Systems: Human Factors Issues*. NATO ASI Series/Computer and Systems Sciences Vol. 110 (New York: Springer) 401–16.

Westrum, R. (1999), *Sidewinder: Creative Missile Design at China Lake* (Annapolis: Naval Institute Press).

Westrum, R. (2006a), "A Typology of Resilience Situations," in E. Hollnagel, D.D. Woods, and N. Leveson (eds), *Resilience Engineering. Concepts and Precepts* (Aldershot, UK: Ashgate).

Westrum, R. (2006b), "Katrina: When Resilience Fails," in *Proceedings of the 2nd Symposium on Resilience Engineering*. Juan les Pins, France, 8-10 November.

Westrum, R., and Adamski, A.J. (1999), "Organizational Factors Associated with Safety and Mission Success in Aerospace Environments," in D.J. Garland, J.A. Wise, and D. Hopkin (eds), *Handbook of Aviation Human Factors* (Mahwah, NJ: Laurence Erlbaum Associates).

Wheelwright, J. (1994), *Degrees of Disaster: Prince William Sound: How Nature Reels and Rebounds* (New York: Simon and Schuster).

Wiener, N. (1961), *Cybernetics: Or the Control and Communication in the Animal and the Machine*, 2nd edn (Cambridge, MA: MIT Press).

Wiens, J.A. (1996), "Oil, Seabirds and Science," *BioScience* 46:8, 587–93.

Wilcox, R.K. (1990), *Scream of Eagles: The Creation of Top Gun and the U.S. Air Victory in Vietnam* (New York: Wiley).

Wildavsky, A. (1988), *Searching for Safety* (New Brunswick: Transaction Books).

Winter, S.G. (2003), "Understanding Dynamic Capabilities," *Strategic Management Journal* 24:10, 991–5.

Winters, B.D., Pham, J.C., Hunt, E.A., Guallar, E., Berenholtz, S., and Pronovost, P.J. (2007), "Rapid Response Systems: A Systematic Review," *Critical Care Medicine* 35:5, 1238–43.

Wolpert, D.H., and Macready, W.G. (1997), "No Free Lunch Theorems for Optimization," *IEEE Transactions on Evolutionary Computation* 1:1, 67–82.

Woods, D.D. (1998), "Designs are Hypotheses about How Artifacts Shape Cognition and Collaboration," *Ergonomics* 41:2, 168–73.

Woods, D.D. (2005), "Creating Foresight: Lessons for Resilience from Columbia," in W.H. Starbuck and M. Farjoun (eds), *Organization at the Limit: NASA and the Columbia Disaster* (Malden, MA: Blackwell) 289–308.

Woods, D.D. (2006), "Essential Characteristics of Resilience," in E. Hollnagel, D.D. Woods, and N. Leveson (eds), *Resilience Engineering: Concepts and Precepts* (Aldershot, UK: Ashgate).

Woods, D.D. (forthcoming), "Escaping Failures of Foresight," *Safety Science*.

Woods, D.D. and Cook, R.I. (2004), "Mistaking Error," in B.J. Youngberg and M.J. Hatlie (eds), *The Patient Safety Handbook* (Sundbury, MA: Jones and Bartlett Publishers) 95–108.

Woods, D.D., and Cook, R.I. (2006), "Incidents: Are They Markers of Resilience or Brittleness?" in E. Hollnagel, D.D. Woods, and N. Leveson (eds), *Resilience Engineering: Concepts and Precepts* (Aldershot, UK: Ashgate).

Woods, D.D., Dekker, S.W.A., Cook, R.I., Johannesen, L.L. and Sarter, N.B. (forthcoming), *Behind Human Error*, 2nd edn (Aldershot, UK: Ashgate).

Woods D.D., and Hollnagel, E. (2006), *Joint Cognitive Systems: Patterns in Cognitive Systems Engineering* (New York: Taylor and Francis/CRC Press).

Woods, D.D., and Shattuck, L.G. (2000), "Distant Supervision: Local Action Given the Potential for Surprise," *Cognition Technology and Work* 2:4, 242–5.

Woods, D.D., and Wreathall, J. (2008), "Stress-Strain Plot as a Basis for Assessing System Resilience," in E. Hollnagel, C. Nemeth, and S.W.A. Dekker (eds), *Resilience Engineering Perspectives, Volume 1: Remaining Sensitive to the Possibility of Failure* (Aldershot, UK: Ashgate) 145–161.

Work, P.A., Rodgers, S.M., and Osborne, R. (1999), "Flood Retrofit of Coastal Residential Structures: Outer Banks, North Carolina," *Journal of Water Resources Planning and Management* 125:2, 88–93.

Wreathall, J. (1989), "A Hierarchy of Risk Control Measures, With Some Consideration of 'Unorganizational Accidents'," in *Second World Bank Workshop on Risk Management and Safety Control* (Karlstad, Sweden: The World Bank).

Wreathall, J. (2001), "Systemic Safety Assessment of Production Installations," in *World Congress: Safety of Modern Technical Systems* (Saarbrucken, Germany: TUV-Verlag).

Wreathall, J. (2006), "Properties of Resilient Organizations: An Initial View," in E. Hollnagel, D.D. Woods, and N. Leveson (eds), *Resilience Engineering: Concepts and Precepts* (Aldershot, UK: Ashgate) 275–85.

Wreathall, J. (2007), "Resilience Engineering: Managing the Residual Risks of Risk Assessment." Paper presented at *Open Initiative for Next Generation PSA*. Washington, DC. 3 October.

Wreathall, J., and Merritt, A.C. (2003), "Managing Human Performance in the Modern World: Developments in the US Nuclear Industry," in G. Edkins and P. Pfister (eds), *Innovation and Consolidation in Aviation* (Aldershot, UK: Ashgate).

Wright, P.C., and McCarthy, J. (2003), "Analysis of Procedure Following as Concerned Work," in E. Hollnagel (ed.), *Handbook of Cognitive Task Design* (Mahwah, NJ: Lawrence Erlbaum Associates).

Xia, W., and Lee, G. (2004), "Grasping the Complexity of IS Development Projects," *Communications of the ACM* 47:5, 68–74.

Xiao, Y., Lasome, C., Moss, J., Mackenzie, C.F., and Faraj, S. (2001), "Cognitive Properties of a Whiteboard: A Case Study in a Trauma Centre," in *Proceedings of the Seventh Conference on European Conference on Computer Supported Cooperative Work* (Bonn, GER: Kluwer Academic Publishers) 259–78.

Zahra, S.A., and George, G. (2002), "Absorptive Capacity: A Review, Reconceptualization, and Extension," *Academy of Management Review* 27:2, 185–203.

Zhou, T., Carlson, J.M., and Doyle, J. (2002), "Mutation, Specialization, and Hypersensitivity in Highly Optimized Tolerance," in *Proceedings of the National Academy of Sciences* 99:4, 2049–54.

Zhou, T., Carlson, J.M., and Doyle, J. (2005), "Evolutionary Dynamics and Highly Optimized Tolerance," *Journal of Theoretical Biology* 236:4, 438–47.

# Index

## A

ability 23–6, 28–33, 37–9, 42–3,
 63–4, 68–70, 73, 83–4, 97–8,
 105, 117, 139–42, 144–9,
 202–3, 260–2, 274–5
accidents 15–16, 26, 40, 129,
 140–1, 150–1, 155, 165, 190,
 287
adaptations 104–5, 136, 184, 232,
 234, 278, 282
adaptive capacities 19, 28, 34,
 95–6, 102, 105, 107, 109–15,
 195, 228
adaptive systems 105, 109–13
agility 5, 62, 65, 75–8, 82, 86–9,
 282, 284, 286
aid 31, 39–40, 47–8, 54, 60, 88,
 257–8, 269
aircraft 174, 179, 181, 183, 185,
 229–32, 251, 254
alternative resources (AR) 209,
 218–21
AR *see* alternative resources
artefacts 249–50, 252, 255, 257–8
astronauts 157, 159–60, 169–70
attention 27, 29, 37–8, 147, 152,
 163, 170, 193–4, 197, 225,
 253, 259
attributes 62, 88, 90–1, 96–7
authority 101, 181, 185, 240,
 243
aviation 3, 13–14, 18–19, 33, 141,
 172, 233–4, 284, 290, 299

## B

balance 67, 81, 113, 126, 129,
 157
behavioural resilience 64, 66, 68,
 70–1, 74, 87
behaviours 27–8, 52–3, 55, 65–72,
 74, 76, 80, 87, 98, 203, 206,
 225
boundaries 31, 99–100, 102,
 202–3, 205, 219–21, 225
bridges 39, 41, 71, 73, 164, 255
brittleness 19, 34, 96, 101, 108,
 110–11, 113–15, 184–5, 275,
 278, 294, 298

## C

CA (Chemical Advisor) 207, 214,
 219, 222–4, 282, 287, 289,
 291–4, 297
capabilities 14, 62–5, 67, 75–6, 78,
 83, 88, 90, 101, 117, 142, 162,
 165, 287, 295
capacity 27–8, 42–3, 56, 81, 86, 90,
 104–5, 108, 110–11, 115, 173,
 221, 228, 258–9, 274
 buffering 203–4, 218, 225–6
challenges 9, 23–4, 29, 32, 34,
 37–8, 45, 68, 73, 247, 258–9,
 274–5, 293
Chemical Advisor *see* CA
city 39–41, 49–50, 59–60, 169, 277,
 284

For Product Safety Concerns and Information please contact our EU
representative GPSR@taylorandfrancis.com
Taylor & Francis Verlag GmbH, Kaufingerstraße 24, 80331 München, Germany

www.ingramcontent.com/pod-product-compliance
Ingram Content Group UK Ltd.
Pitfield, Milton Keynes, MK11 3LW, UK
UKHW021620240425
457818UK00018B/661